Biofertisers for
Sustainable Agriculture

Biofertisers for Sustainable Agriculture

Vinod Kumar Jain

Oxford Book Company
Jaipur ▪ New Delhi

ıed By

:ord Book Company

Off. "Bony Residency", Gate No. 2
ilak Public School, Vishweswariya Nagar,
ɔura Road, Jaipur - 302018, India
41 - 2761280, Telefax: 2761381
oxfordbookjpr@gmail.com
te - oxfordbookcompany.com

Office: 102, 1st Floor, Satyam House,
3, Ansari Road, Darya Ganj,
)elhi - 110002, India
1-45652440

ıted by
INTERNATIONAL PRESS LLP
: keyainternationalpressllp@gmail.com

ıed in India, 2026

9788189473778

₹ 1900/-

:t by Shivangi Computers, Jaipur

ɪ and bound in India by Replika Press Pvt. Ltd.

Preface

Sustainable agriculture, with its emphasis upon the exclusion of synthetic fertilisers, pesticides and chemical agents, and the inclusion of eco-friendly farm practices, is a practice which is increasingly finding favour worldwide. We live in an age wherein a growing need has been felt to grow food in sufficient amount to feed the ever growing world population, without comprising on either nutritional quality or ecological balance. In such a scenario, sustainable agriculture, albeit fraught with various hurdles and challenges, presents a highly viable option. Biofertilisers, environmentally sustainable and pollution-free, are an important input to this mode of agricultural practice.

This book has been designed as an introductory manual which seeks to acquaint readers with the essentials of biofertiliser use in sustainable agriculture practice. The aim is to unravel what entails biofertilisers, their types, nutritional qualities, their use and the methodology of producing them. In addition, the book takes care to incorporate critical perspectives which look into biofertiliser use as per its advantages, the challenges surrounding its use and the future offered by it.

Binod Kumar Jain

Contents

1

Principles of Sustainable Agriculture

Sustainable agriculture has been practiced for many decades and encompasses a tremendous number of different approaches described by many different names. To this point, most of these approaches have largely been limited to the substitution of environmentally. More significant advances can be expected, however, as a result of developments in the science and art of agroecosystem design and management. Many of the approaches in conventional agriculture would fall into the "efficiency" category. They demonstrate a reduction in resource use and associated negative environmental impact, and in many cases a reduction in input expenses for the farmer. They represent, however, only an initial step towards a truly sustainable system.

Efforts to substitute safe products and practices are also gaining popularity.Despite the reduced negative environmental damage associated with them, they remain problematic. Botanical pesticides also kill beneficial organisms, the release of biocontrols does not address the question of why pest outbreaks occur, dependence on imported fertiliser materials makes the system vulnerable to supply disruptions and excessive cultivation to control weeds is detrimental to the soil.

The systems that focus on redesign of the farm are the most sophisticated and generally the most environmentally and economically sustainable, especially over the long term. These farm systems recycle resources to the greatest extent possible, meaning that little is wasted, few pollutants are generated, and input costs are reduced substantially. For example, chicken and orchard operations have been successfully integrated. The manure is used as a fertiliser, the chickens eat pests that attack the fruit, the feed bill for the chickens is greatly reduced, and the eggs and/or meat can be consumed or sold. Three to seven year crop rotations can be designed that minimise tillage, use legumes and green manures to maintain soil fertility, prevent pest and disease outbreaks, and provide a diverse diet for livestock.

Pigs and goats can be used to renovate wooded lands in preparation for sheep pasture. The pigs and goats replace the petrochemical energy that would be consumed in machines, herbicides and fertilisers. All these practices involve redesigning the farm. As in conventional agricultural systems, the success of sustainable approaches is very dependent on the skills and attitudes of the producers. The degree to which different models of such farms are sustainable is very variable, and is dependent on the physical resources of the farmer, and the degree deficiencies in support farm, the talents and commitment of the of support available. The current from government, universities, and agricultural professionals means that farmers must often rely on their own talents and commitment.

ECONOMICS OF SUSTAINABLE AGRICULTURE

Once a farm has passed through the transition period, it is likely to be at least as profitable, if not more so, than under conventional practices. The principal reason for this is the decline in input costs, to, on average, 1/3 below conventional costs. Yields commonly decline in some crops but increase in others, especially during dry years, when the better water holding capacity of the sustainably managed soil translates into a production advantage. Existing literature suggests that, when yields of commonly grown crops are averaged, there is an overall 10′ yield decline compared to yields in conventional production systems.

Total yield of all products on any individual farm, however, that takes place, and the addition of complementary animal and crop production systems. If a premium price can be obtained for the products, then the profitability of the operation may exceed that under conventional operation. Returns to labour are often lower, either because yields are somewhat lower or because labour requirements increase. However, labour needs are usually more spread out over the year. Work may also be more rewarding as a result of diversification and reduced risks to health. On a regional or national level, it appears that more widespread adoption of sustainable agriculture can meet many of the government's stated policy objectives for agriculture.

Studies done to date indicate that total net farm income would increase, government subsidy payments could decline, environmental damage would decline, food quality would improve, and rural employment possibilities rise. Potential problems that could develop include a decline in export of major commodities such as wheat (due to diversification of production), dislocations in the farm input supply sector, and a shortage of skilled labour, manure and sources of potassium.

CERTIFIED ORGANIC FOOD AND FARMING

Some producers have chosen to employ sustainable production practices that will permit them to market their produce as "certified organic". There are no national estimates yet of what percentage of production from sustainable practices is marketed in this way, but in Quebec, the figure is estimated at about 30%.

Use of the label "certified organic" developed as a way to assure consumers that the food they are eating is, in fast, grown according to the practices that are commonly associated with the word "organic". Those involved in promoting "organic" food have seen what happened to the "natural" food market. Because "natural" was not clearly described it was easy for the work to be co-opted and to be used to describe almost any kind of food product or process. The term "organic", first used extensively in the USA by J.I. Rodale. founder of Rodale Press, has continued to be used because it is the term most understood by consumers in the market place. Attempts to use other words that more accurately describe sustainable agriculture practices have failed to become established, principally because they had no meaning in the marketplace.

The certification process is useful in our food economy because consumers usually do not know the farmer whose products they are buying. Many organic growers are involved in interprovincial and international trade. In some countries, such as Japan, certification has not been as important a development. The Japanese sustainable agriculture movement has instead focused on bringing consumers and producers closer together by creating consumer producer cooperatives and buying groups, thereby reducing the need for certification. In this kind of system, consumers may even be involved in farm management decisions. This approach is also being practiced in a few places in North America.

Certification standards have been developed for most provinces in Canada. Although they all have a common base, they often differ in details, depending on the ecological and economic conditions that exist in the region. Although based on "agroecological" principles, each set of standards is in fact a compromise between the ideal situation and the state of development of sustainable practices in each region. For example, in some standards, certain fertilisers and pesticides are permitted even though they may have detrimental effects on beneficial soil organisms, natural pest control agents and wildlife. In many cases, our understanding of the ecology of a pest or production system has not advanced to the point where we can assure a productive and profitable system without using such products over the short term.

ECOLOGY AS A TOOL TO UNDERSTAND SUSTAINABLE AGRICULTURE

It is common to hear a farmer say, "I don't grow my crops this way because it's organic, I do it because it makes good agronomic sense". What has become increasingly clear to a growing number of agricultural professionals in the last few years is that good agronomy is based on an understanding of ecology. An "agroecological" approach is used increasingly by agricultural professionals to analyse the success of sustainable farming systems, and to identify ways of improving the productivity profitability, and resource efficiency of them.

Agroecology is a relatively new scientific approach and has not yet attained the stature of other agricultural disciplines. As a result, most of today's agriculture: professionals have little formal training in this area, or' soil or even crop and livestock ecology. This situation : changing rapidly, particularly in the USA and Europe.

Many schools now offer undergraduate and graduate programmes in ecological agriculture. For example, the core undergraduate programme of the University of Maine is the Sustainable Agriculture Programme. In Quebec, the greatest interest in providing training in this area is coming from the Commissions Scolaires, the ITAs and the CEGEPs. Macdonald College will be offering a 24-course Minor in Ecological Agriculture starting in September 1989. Laval is beginning a programme entitled Agriculture and Environment. Guelph is developing a course, and a number of community colleges across the country are also investigating possible offerings in this area. Some governments, including MAPAQ, are preparing training programmes for their agricultural professionals in order to be able to provide services to interested farmers.

TECHNIQUES

Monoculture, a method of growing only one crop at a time in a given field, is a very widespread practice, but there are questions about its sustainability, especially if the same crop is grown every year. Growing a mixture of crops (polyculture) sometimes reduces disease or pest problems. For example, how does growing a corn-bean mixture every year compare with growing corn and bean in alternate years? Cropping systems that include a variety of crops (polyculture and/or rotation) may also replenish nitrogen (if legumes are included) and may also use resources such as sunlight, water, or nutrients more efficiently.

Replacing a natural ecosystem with a few specifically chosen plant varieties as is done in farming results in an artificial ecosystem that lacks the genetic diversity found in wildlife and is thus more susceptible to widespread disease. The Great Irish Famine is a well-known example of the dangers of monoculture.

Many scientists, farmers, and businesses have debated how to make agriculture farming sustainable. One of the many practices includes growing a diverse number of perennial crops in a single field, each of which would grow in separate season so as not to compete with each other for natural resources. This system would result in increased resistance to diseases and decreased effects of erosion and loss of nutrients in soil. Nitrogen fixation from legumes, for example, used in conjunction with plants that rely on nitrate from soil for growth, helps to allow the land to be reused annually. Legumes will grow for a season and replenish the soil with ammonium and nitrate, and the next season other plants can be seeded and grown in the field in preparation for harvest.

In practice, there is no single approach to sustainable agriculture, as the precise goals and methods must be adapted to each individual case. There may be some techniques of farming that are inherently in conflict with the concept of sustainability, but there is widespread misunderstanding on impacts of some practices. For example, the slash-and-burn techniques that are the characteristic feature of shifting cultivators are often cited as inherently destructive, yet slash-and-burn cultivation has been practiced in the Amazon for at least 6000 years; serious deforestation did not begin until the 1970s, largely as the result of Brazilian government programmes and policies.

There are also many ways to practice sustainable animal husbandry. Some of the key tools to grazing management include fencing off the grazing area into smaller areas called paddocks, lowering stock density, and moving the stock between paddocks frequently. Several attemps have been made to produce an artificial meat, using isolated tissues to produce it in vitro; Jason Matheny's work on this topic, whichin the New Harvest project, is one of the most commented.

OFF-FARM IMPACTS

What if a farm is able to "produce perpetually", yet has negative effects on environmental quality elsewhere? Most people concerned with sustainability take a global view, so they try to avoid negative off-farm impacts. For example, over-application of synthetic fertiliser or animal manures can pollute nearby rivers and coastal waters. On the other hand, if crop yields are too low, because of soil exhaustion of nutrients or reduced ability to retain water, farmers would need to access new lands for agriculture, leading to the decimation of the rainforest, draining wetlands, etc.

We must also consider the impact of sustainability on overall production. If the human population is to continue its growth, it will need food and fibber to do so. The United Nations estimates the world will inhabit 9.3 billion people by 2050, which will necessitate a dramatic increase in production capabilities. The increased production will likely come from one of two different ways. First, you can either break out virgin land to grow crops, though concerns over global warming make this option unsavory. Second, you can increase yields on existing land, which will likely require adopting technologies many sustainable advocates are opposed to, principally Genetically modified organism crops.

Many advocates of sustainable agriculture think that organic agriculture is the only system which can be sustained over the long-term. However, organic production methods, especially in transition, yield less than their conventional counterparts. While there is evidence which gives organic an advantage during periods of drought one must be careful not to place too much emphasis on these figures.

While drought is a real threat to agriculture production, it is hardly the norm. Even when drought strikes, stockpiles help mitigate food security concerns. Regardless of reduced yields under optimal conditions, it should be noted that during periods of prolonged drought as predicted by global warming scientists, organic production methods should be considered as a way to adapt to a changing climate.

URBAN PLANNING

There has been considerable debate about which form of human residential habitat may be a better social form for sustainable agriculture. Generally, it is thought that village communities can improve sustainability in that such communities tend to provide a cooperative environment that supports farming.

Many environmentalists pushing for increased population density to preserve agricultural land point out that urban sprawl is less sustainable and more damaging to the environment than living in the cities where cars are not needed because food and other necessities are within walking distance. However, others have theorised that sustainable ecocities, or ecovillages which combine habitation and farming with close proximity between producers and consumers, may provide greater sustainability. The use of available city space for cooperative food production is another way to achieve greater sustainability.

One of the latest ideas in achieving sustainable agricultural involves shifting the production of food plants from major factory farming operations to large, urban, technical facilities called vertical farms. The advantages of vertical farming include year-round production, isolation from pests and diseases, controllable resource recycling, and on-site production that eliminates the need for transportation costs.

While a vertical farm has yet to become a reality, the idea is gaining momentum among those who believe that current sustainable farming methods will be insufficient to provide for a growing global population. For vertical farming to become a reality, billions of dollars in tax credits and subsidies will need to be made available to the operation. It may be difficult to justify spending billions of dollars on a vertical farm that will only feed 50,000 people when agriculture land remains abundant.

IMPLEMENTING SUSTAINABLE AGRICULTURE

Until World War II, increases in agricultural production throughout the world, including North America, primarily involved bringing more land into cultivation. However, during the last 50 years, the rate at which new land has been brought into cultivation has declined sharply. New cropland worldwide expanded at only 0.3% per year in the 1970s compared with a rate of 1% in the 1950s. Future increases in agricultural production will have to come primarily from increasing output per unit of land rather than increasing the area cultivated. Remarkable increases in agricultural productivity have occurred in many parts of the world, including North America, during the last two to three decades. However, concerns are increasing as to whether these gains can be maintained in a sustainable manner.

A major reason for these concerns is the increasing energy cost per bushel of crop produced in the face of declining fossil energy reserves, meaning that the output/input energy ratios are becoming progressively less favourable. Conventional modern farming has had a steadily declining output/input ratio since the 1940's, going from 3.0 to less than 1.8 by 1980. Thus, land-use strategies using concentrated, finite, energy inputs will not be sustainable in the long run. In light of such concerns, the critical issue is whether agricultural production per acre and per kilocalorie can be increased simultaneously.

Common Impressions about Sustainable Agriculture

Many advocates of sustainable agriculture would not agree and argue that success in moving toward sustainable agriculture depends on using the limited nonrenewable resources as sparingly as possible and getting maximum return from their application by using the biological cycles that exist in nature and are largely ignored in present-day agriculture.

The following impressions have contributed to a lack of support for sustainable agriculture.

— It is similar to the type of agriculture that was practiced in the early 1900's and involved purchasing few inputs and marketing little of the products produced. Such agricultural systems were found commonly where abundant agricultural land was available. Good examples are the crop rotations practiced by the early pioneers in various parts of the United States.

— Farmers value their freedom of action, and the recent interest in sustainable agriculture by governmental agencies has been viewed negatively rather than positively by some. Major governmental thrust in this direction has been understood, perhaps unjustifiably, as involving restrictions on farm practices rather than incentives and/or voluntary compliance. Fortunately, some of the provisions in the 1996 Farm Bill appear to be moving toward softening this philosophy.

— Until very recently, those directly associated with sustainable agriculture were viewed as hippies who were out of touch with so-called modern/progressive agriculture. The history of such attitudes goes back to the 1970s when the alternative agriculture movement, which often was used interchangeably with sustainable agriculture, started emerging. Fortunately, this attitude of "them" and "us" has broken down to some extent in recent years.

Although aspects of these impressions may still exist among some people, they are no longer major barriers to constructive dialogue on sustainable agriculture. Most people now recognise the need to be directly concerned about the sustainability of agriculture, although opinions differ on how it could or should be achieved. Thus, another issue is agreeing on a definition. Many definitions of sustainable agriculture have been proposed, but one of the first to be adopted in the US was published by the American Society of Agronomy:

> "A sustainable agriculture is one that, over the long term, enhances environmental quality and the resource base on which agriculture depends; provides for basic human food and fibber needs; is economically viable; and enhances the quality of life for farmers and society as a whole."

In the following year, the United States Congress also defined sustainable agriculture in the 1990 Farm Bill. Under that law, the term sustainable agriculture means an integrated system of plant and animal production practices having a site-specific application that over the long term will:

— Satisfy human food and fibber needs.
— Enhance environmental quality and the natural resource base upon which the agricultural economy depends.
— Make the most efficient use of nonrenewable resources and on-farm resources and integrate, where appropriate, natural biological cycles and controls.
— Sustain the economic viability of farm operations.
— Enhance the quality of life for farmers and society as a whole.

Thus, the official, or legal, definition has five parts, emphasizing productivity, environmental quality, efficient use of nonrenewable resources, economic viability, and quality of life. Under this definition, a farm that emphasizes short-run profit, but sacrifices environmental quality, would not be sustainable in the long run. From the other end, pursuing environmental quality without ensuring viability of short-run returns also would be unsustainable. A farm that is very productive but uses large quantities of a nonrenewable resource, such as fossil fuel or a non-rechargeable aquifer, to achieve and maintain that productivity would not be considered sustainable in the long run.

In 1996, focus group discussions were held with some Kansas farmers who are practising sustainable agriculture. For them, agricultural sustainability consists of three components: ecological, economic, and social/institutional. In order to achieve these, they recognised three important "processes": full use of the natural biological cycles, reducing as much as possible the use of purchased inputs, and reducing the frantic work schedules of many farm families. Thus, their perception of sustainable agriculture is similar to the more formal definitions mentioned earlier.

Sustainable Agriculture and Other Terms

Other terms sometimes have been associated with sustainable agriculture.

Some people believe that the terms *organic* and sustainable are the same. In fact, the word organic refers to particular farming practices that have been followed and certified by a third-party inspector. This organic certification is found on the product in the marketplace, indicating to consumers that it has been grown in a specific way. Briefly, organic guidelines have been written in an attempt to be as sustainable as possible, while assuring the consumer that the amount of pesticide residue on the food product is minimised.

Organic guidelines require that a crop rotation and soil-building practices be used; that pest control rely primarily on cultural, mechanical, and biological controls; and that pesticides used as a last resort are from plants and are short-lived in the environment. Soil amendments should come from renewable resources as much as possible and only nonrefined mined products used, when manure, composts, or other sources of nutrients are not available.

The term *low input* also has been associated with sustainable and, in fact, the first USDA funding for sustainable agriculture research in the 1985 Farm Bill used the acronym LISA for 'low input sustainable agriculture'. The term LISA was somewhat misleading and indeed unfortunate. For some it implied that farmers should starve their crops, let the weeds choke them out, and let insects clean up what was left. In fact, the term low-input referred to purchasing fewer off-farm inputs (usually fertilisers and pesticides), while increasing on-farm inputs. Thus, a more accurate term would be *different input* or *low external input* rather than low-input.

In Europe (e.g., the Netherlands), the term *biological* often refers to *organic* farming, whereas the term *ecological* refers to organic plus environmental considerations such as on-farm wildlife management. The term *integrated* is used both in the US and in Europe, generally to refer to a system of *integrated pest (or crop) management*, which can demonstrate some aspects of sustainable agriculture but may or may not fit the other parts of the sustainability definitions given above.

Sustainable Agriculture Myths

As indicated earlier, even today, sustainable agriculture sometimes gets bad press. Sometimes this is due to impressions based on what we term myths or misconceptions about sustainable agriculture. Three commonly held myths are the following:

Myth 1: Sustainable agriculture means going back to what our grandparents did or involves rejecting appropriate technology. In fact, sustainable agriculture seeks to combine some of the wisdom of past practices, like crop rotation and green manure crops, with careful use of current technology, including equipment that leaves residue on the surface, modern pest-resistant crop varieties, and the vast array of information technologies now available.

Myth 2: Adoption of sustainable agriculture will inevitably involve losing money. Studies from both a real-life case study approach and from economic "what if" type models have documented that sustainable agriculture can be as profitable in the short run as conventional or less sustainable alternatives.

The question then arises as to why practices consistent with sustainable agriculture are not adopted more widely. One reason has been that many production questions about sustainable agriculture are unanswered, partly because of continuing low levels of research funding available for directly addressing sustainable agriculture issues. Other factors that limit acceptance have included the base acres requirement in the pre-1996 Farm Bills and the amount of learning required for nonchemical pest control and biological changes that occur when one begins using manures and legumes on soil that has had no recent, significant, organic amendment.

Economic survival during the transition period is crucial. Obvious biases and hidden biases in the current economic system often have made conventional agriculture more profitable for individual producers. This has been supported by the development of economies of scale that encourage specialisation and, hence, larger farms characteristic of conventional agriculture. The economic pressures encouraging the development of larger farms, according to Smith, have resulted from producers losing control of the input and processing/marketing of the product. Thus, it is not surprising to find sustainable agriculture practitioners establishing greater control of aspects of production and marketing.

Certainly, the economies of scale associated with specialisation (resulting in short-term financial advantages) mean greater challenges for large farms to adopt sustainable agricultural practices. Also, the adoption of sustainable agriculture practices is viewed by practitioners as fairly management intensive. As Dan Nagengast, Executive Director of the Kansas Rural Center, has observed: "though farming sustainably can be less expensive, and more profitable, it will require intensive management, which is another way of saying, it requires an engaged mind."

Myth 3: Sustainable agriculture increases soil erosion because of greater use of tillage. Sustainable agriculture practitioners must meet conservation compliance guidelines but can emphasise some methods of soil conservation over others. For example, cropping under no-till systems generally does not work well for organic agriculture farmers, because of the inability to reduce herbicide use. On the other hand, a tillage rotation system, where no-till can be alternated with reduced-till and conventional-till practices, is feasible. Another alternative is ridge tillage, which potentially can reduce both tillage and chemical use.

Sustainable agriculture practices also involve having a higher percentage of crop land protected with a living cover crop during winter months, as well as some rotation with perennial crops, sods, or alfalfa that help hold the soil and improve the water infiltration rate, thus reducing run-off when the field is put back into an annual crop. Row crops can be cultivated for weed control, but rapid canopy closure and overseeding with fall cover crops help reduce erosion potential during the rest of the cropping season.

Although ultimately the decision as to whether or not to practice sustainable agriculture is the made by the farmers and their families, the ease and practicality of doing so are affected heavily by a number of factors, some of which they can influence, but some of which are completely out of their control. Generally, progress made toward sustainable agriculture is determined by what happens at five different levels: international, national, community, farm, and field. Another level is involved that cannot be fitted easily into the above classification. This is the watershed level, which may be at the field, farm, or community level or some combination of the three.

Thus, agricultural sustainability not only is a difficult concept to define but also is difficult to implement and monitor/measure. This complexity is illustrated in Table 1 by showing the expected interactions among the three components of sustainability and the five levels of influence. Although sustainability tends to be locational or site specific (at the field, farm, and community levels), it is very much influenced by:

Table 1: Interacting Components of Sustainability

Levels Influencing Sustainability	*Components of Sustainability* *Ecological*	*Economic*	*Social/Institutional*
International	Secondary	Secondary	Secondary
National	Secondary	Secondary	Primary
Community	Secondary	Primary	Primary
Farm	Primary	Primary	Primary
Field	Primary	Secondary	Secondary

1. *What happens at the higher levels:* National policies such as in the 1996 Farm Bill will have a great influence on ecological and economic sustainability at the field/farm levels. Other policies at that level related to social/institutional issues also can have major effects on the viability/welfare of communities and, hence, on quality of life. International markets and influences (particularly in smaller countries) are increasingly affecting what happens at the lower levels. Such influences tend to be relatively greater in countries that are poor (low income) and/or where agricultural production is influenced heavily by the export market. Thus, it is necessary to understand the interaction between these levels, because "each level finds its explanations of mechanism in the levels below, and its significance in the levels above".
2. *Interactions among the sustainability components:* In the focus group discussions referred to above, some of the farmers indicated that those who were in conventional agriculture were often on an economic treadmill and hence had little time to consider ecological sustainability issues. They also had to make compromises concerning quality of life because of having to work very long hours. In fact, the prevailing

attitude among the farmers was that all three components of sustainability had to be pursued at the same time, if progress was to be achieved.

SUSTAINABLE AGRICULTURE ISSUES

Given the complexity of sustainable agriculture in terms of both definitions and interactions, it is not easy for farmers to implement and monitor or measure it. A lot of time and space could be spent expanding on these points, but three major issues can summarise much of our thinking relating to implementing sustainable agriculture.

1. The work required to ensure agricultural sustainability, although influenced to some extent by outside forces, is mainly field or farm specific (Table 1). This implies heavy dependence on decision-making by farmers and their families, which has led those promoting sustainable agriculture to advocate *whole farm planning (WFP)*. Farmers and their families go beyond looking at only the economics of their farm "business" and include issues relating to ecological sustainability and quality of life in their planning process. This, in turn, requires much more of a systems perspective in asking the correct questions rather than relying on obtaining recipe-type solutions. Such solutions are most applicable in response to narrowly defined technical and economic problems that occur in conventional systems, where the farm tends to be viewed as a business for ensuring short-run economic gain. A shift to whole farm planning in the approach to farming and life is a major challenge that takes time and energy and often involves questioning conventional wisdom and the status quo. This shift also implies the need for greater empowerment of farmers and their families, something that is increasingly novel with our current economic system and society!
2. The multifaceted nature of sustainable agriculture,

with three interdependent and interactive components, makes measuring and monitoring it extremely difficult. Not surprisingly, a number of indicators are currently emerging that measure the different components. However, at least three major challenges remain.

— The measures currently available generally fall short in terms of assessing the interactions and interdependencies among the three components and the trade-offs of pursuing one component at the expense of another.

— Many of the measures or indicators currently available are not particularly useful to farmers or are too time-consuming to measure in their day-to-day work, making it difficult for them and their families to monitor progress in terms of agricultural sustainability. This is particularly regrettable because many of the issues relating to sustainable agriculture are location or situation specific.

— Most indicators show progress or no progress towards specific components of sustainability, but they fall short in terms of helping to determine cause/effect relationships to help assess current problems and provide ideas on what needs to be done to ensure continued progress towards sustainability. An additional complication is that some strategies relating to sustainable agriculture require 5 to 10 years (e.g., a full crop rotation) of implementation before they result in visible or measurable signs of payoff.

3. The current, conventional, economic system places a great deal of reliance on prices to send the correct signals as far as society is concerned. Sometimes curbs are put in place to modify such prices. However, these generally have not favoured or advanced the cause of sustainable agriculture, as defined above. Concerns about how money and market prices send the right

signals, in relation to sustainable agriculture, are the following:

— Wealth and the related issue of happiness as viewed in sustainable agriculture consist of more than money, which is the sole measure in conventional economic systems. Therefore, using money as a yardstick or evaluation criterion for sustainable agriculture has deficiencies. For example, it does not ensure ecological sustainability (part of the wealth of future generations), community sustainability, or a certain quality of life (which consists of more than buying happiness).

— Market prices are concerned only with the product that enters the marketplace, whereas quality of life involves more than what can be expressed in the marketplace. The market does not always handle interdependencies very well. For example, cheaper mechanical equipment displaces locally hired labour and reduces community sustainability. Another example of the failure of the market system to address interdependencies relates to nonfarm problems that emerge as a result of a farming practice (e.g., farmers using excessive chemicals to help increase their own short-run production/income and polluting water accessed by others). Also, the market often excessively discounts the future, either implicitly or explicitly (through specific governmental policies).

Therefore, gross or net income is not necessarily a good indicator of "success," and costs do not necessarily reflect the "cheapest" way of producing products, especially in the long run when non-renewable resources are being used and/or the land is being "mined" to produce current profits.

Another way of viewing the above discussion is in terms of three types of "capital". Financial or economic capital is reflected in the marketplace. However, the marketplace often does a poor job of reflecting the other two types of capital: natural or ecological and social. Natural or ecological capital includes measures of agroecosystem biodiversity, soil and water quality, and energy source. Social capital includes such things as accumulation of skills, training, education, social connectedness, sharing, and population stability. Currently, the primary "feedback" in determining agroecosystem input use is market force or economics. In an ideal world, the other two sources of capital also would have effective feedbacks in influencing such input decisions.

Therefore, advocates of sustainable agriculture can argue that the current excessive use of nonrenewable resources that will not be available for future generations inhibits the attainment of sustainability. Sustainable agriculture means less emphasis on extractive methods (and the use of nonrenewable resources) and greater reliance on renewable methods and enhancing the resource base for future generations (that is, exploiting useful biological cycles, thereby saving dollars on externally purchased inputs). This goes against the conventional move towards specialisation and economies of scale, which is encouraged by the current economic/pricing system.

Sustainable agriculture can survive in the current economic climate by working with nature's biological cycles; reducing expenditure for purchased inputs, relying on what Savoury calls solar dollars (income generated through human creativity, labour, and constant sources of energy, especially the sun); and finding niche markets for the products produced. However, given the overwhelming power/influence of the prevailing economic system, popular acceptance of sustainable agriculture in its widest sense will require governmental policies that support farmers not only in the transition to sustainability but in maintaining that commitment once it has been attained. This does not necessarily mean subsidising of farmers per se but rather implementation of policies that are important to ensuring the vitality and viability of society (i.e., families and communities) today and in the future.

TOWARDS A SUSTAINABLE AGRICULTURE

Sustainable agriculture is an important element of the overall effort to make human activities compatible with the demands of the earth's ecosystem. Thus, an understanding of the different approaches to ecological agriculture is necessary if we want to utilise the planet's resources wisely. While sustainable agriculture is based on long-term goals and not a specific set of farming practices, it is usually accompanied by a reduction of purchased inputs in favour of managing on-farm resources.

Low-input farming is based on a reduction—but not necessarily elimination—of chemical fertilisers, insecticides, and herbicides. Farmers are adopting these practices primarily to reduce costs, but also because they want to minimise impact on the environment or because they perceive future pesticide regulations. In a search for information on how to farm with fewer chemicals, it is helpful to examine alternative farming systems in existence that largely exclude chemicals in favour of biological farming practices. Experiences of producers who've successfully practiced these methods are valuable to farmers considering a transition to low-input sustainable agriculture.

Alternative Farming Systems

There are four established approaches to alternative farming in the U.S. A common thread in all four schools is an emphasis on biological systems to supply fertility and pest control rather than chemical inputs. Organic farming is the most widely recognised alternative farming system. Modern organic farming evolved as an alternative to chemical agriculture in the 1940s, largely in response to the publications of J.I. Rodale in the U.S., Lady Eve Balfour in England, and Sir Albert Howard in India.

In 1980, U.S.D.A. released a landmark report titled Report and Recommendations on Organic Farming in which organic farming was defined as such:

> Organic farming is a production system which avoids or largely excludes the use of synthetically compounded fertilisers, pesticides, growth regulators, and livestock feed additives. To the maximum extent feasible, organic farming systems rely upon crop rotations, crop residues, animal manures, legumes, green manures, off-farm organic wastes, mechanical cultivation, mineral-bearing rocks, and aspects of biological pest control to maintain soil productivity and tilth, to supply plant nutrients, and to control insects, weeds, and other pests.

In the 70s and 80s, organic certification of farms emerged as a marketing tool to insure foods produced organically met specified standards of production. The Organic Foods Production Act, included in the 1990 Farm Bill, enabled USDA to develop a national programme of universal standards, certification accreditation, and food labelling. Implementation, initially scheduled for October of 1993, was delayed due to lack of funding and complexity of issues and is anticipated to take effect in 1995. Biodynamic farming evolved in Europe in the 1920s following lectures on agriculture by the Austrian anthropologists Rudolf Steiner.

Biodynamic farming parallels organic farming in many ways but places greater emphasis on the integration of animals to create a closed nutrient cycle, effect of crop planting dates in relation to the calendar, and awareness of spiritual forces in nature. A unique feature of this system is the use of eight specific preparations derived from cow manure, silica, and herbal extracts to treat compost piles, soils, and crops. Demeter (tm) is a certification programme for food and feed produced by strictly biodynamic farming methods. The Community Supported Agriculture (CSA) marketing programmes, gathering popularity as an innovative method of subscription farming, were largely introduced into the U.S. by the biodynamic movement.

An article on soil quality and financial performance of biodynamic and conventional farms in New Zealand in the April 16, 1993 issue of Science. In a comparison of 16 adjacent farms, the biodynamic farms exhibited superior soil physical, biological, and chemical properties and were just as financially viable as their counterparts. "Biological" farming has become synonymous with farmers using the Reams fertility system as the basis for crop production. Eco-agriculture is the term used to describe this system by the monthly Acres, U.S.A. The Reams system is based on the LaMotte-Morgan soil test and the use of rock phosphate, calcium carbonate, and compost to achieve nutrient ratios of 7:1 calcium to magnesium, 2:1 phosphorus to potassium, and so on. "Biological" farming allows the use of selected chemical fertilisers and adopts low-input approaches to use of herbicides and insecticides.

Diagnostic instruments to monitor plant and soil conditions are frequently used in "Biological" farming; these include refractometers to monitor sugar content (Brix) in plant tissue sap; electrical conductivity meters to monitor ERGS (or energy released per gram of soil); ORPS meters (or oxygen reduction potential of soil); and radionics. Based on data gathered, foliar sprays containing biostimulants and soluble nutrients are applied.

The Pandol Brothers, a large commercial fruit and vegetable operation in California, reduced their annual pesticide bill from $500,000 to $50,000 per year after adopting a "Biological" fertility programme. Nature Farming was developed in Japan in the 1930s by Mokichi Okada, who later formed the Mokichi Okada Association

(MOA). Nature Farming parallels organic farming in many ways but includes special emphasis on soil health through composts rather than organic fertilisers, when possible. Kyusei Nature Farming, a branch group, emphasizes use of microbial preparations in addition to traditional Nature Farming.

Nature Farming is most active in the Pacific rim, including California and Hawaii. Since the late 1980s, Nature Farming has gained wider recognition in the United States through the coordinated efforts of MOA and the Rodale Institute in the formation of the World Sustainable Agriculture Association (WSAA). The WSAA and MOA sponsor annual conferences on Nature Farming and sustainable agriculture. Kyusei Nature Farming conducts on-farm research in California. In addition to these methods-based approaches to sustainable farming, regenerative agriculture and permaculture are widely recognised in the U.S. and abroad. However, these latter systems, like sustainable agriculture, are more conceptually oriented than methods-based.

Regenerative agriculture became the preferred term of the Rodale Institute in the late 1970s and 80s under the direction of Robert Rodale. Regenerative agriculture builds on nature's own inherent capacity to cope with pests, enhance soil fertility, and increase productivity. It implies a continuing ability to re- create the resources that the system requires. In practice, regenerative agriculture uses low-input and organic farming systems as a framework to achieve these goals. Permaculture is a contraction of "permanent agriculture" and was coined by Bill Mollison, an Australian forest ecologist, in 1978.

Permaculture is concerned with designing ecological human habitats and food production systems, and follows specific guidelines and principles in the design of these systems. To the extent that permaculture is not a production system, per se, but rather a land use planning philosophy, it is not limited to a specific method of production. Thus, practically any site- specific ecological farming system is amenable to permaculture. A common thread among all six schools is an opposing world view to the industrial model of agriculture. These competing paradigms were summarised in "Conventional Versus Alternative Agriculture: The Paradigmatic Roots of the Debate" as:

— centralisation vs. decentralisation
— dependence vs. independence
— competition vs. community
— dominance of nature vs. harmony with nature
— specialisation vs. diversity
— exploitation vs. restraint

These objectives obviously have more to do with societal and economic responses to modern industrial agriculture than they do with farming practice A or farming practice B. Nevertheless, they underscore the sometimes contentious debate between sustainable farming advocates and supporters of high-input conventional agriculture.

EVOLUTION

In the 1960s and 70s, a growing environmental agriculture movement evolved in response to increasing soil erosion, pesticide use, and groundwater contamination. Simultaneously, economic conditions for farmers were becoming more stressful and the number of family farms declined. In 1980 Wes Jackson of The Land Institute in Salina, KS, began using the term "sustainable agriculture" to describe an alternative system of agriculture based upon resource conservation and quality of rural life. Through the lobbying efforts of several nonprofit farming organisations, Congress passed legislation in the 1985 Farm Bill that mandated implementation of a low-input sustainable agriculture programme by the Department of Agriculture.

In 1988 U.S.D.A. initiated the Low-Input Sustainable Agriculture research and education programme, or LISA. In 1991 the name of this programme was changed to the Sustainable Agriculture Research and Education Programme, or SARE. Funds made available through the LISA/SARE programmes have resulted in significant additions to landgrant research and extension programmes in the last five years. While sustainable agriculture has become the umbrella under which many of the above-mentioned alternative farming systems fall, it is important to note that sustainable agriculture is really a long-term goal, not a specific set of farming practices. In Sustainable Agriculture in Temperate Zones sustainable agriculture was defined as such:

> Sustainable Agriculture is a philosophy based on human goals and on understanding the long-term impact of our activities on the environment and on other species. Use of this philosophy guides our application of prior experience and the latest scientific advances to create integrated, resource-conserving, equitable farming systems. These systems reduce environmental degradation, maintain agricultural productivity, promote economic viability in both the short and long term and maintain stable rural communities and quality of life.

Three indicators that appear most frequently in a definition of sustainable agriculture are:

— Environmentally sound

— Economically viable

— Socially acceptable

In this context, sustainable agriculture embraces all agricultural systems striving to meet these criteria. Many aspects of modern conventional agriculture are included in sustainable agriculture, just as are many aspects of alternative farming systems. One aspect of modern agriculture receiving a lot of attention in the sustainable agriculture discussion is the use of chemical inputs to supply fertility and pest control. While agriculture chemicals will continue to play an important role in American agriculture, many farmers are looking at alternatives due to environmental, economical, or regulatory reasons. In a transition to farming systems more reliant on biological methods of production, low-input farming serves as an intermediary step.

The term low-input agriculture has been defined as a production activity that uses synthetic fertilisers or pesticides below rates commonly recommended by the Extension Service. It does not mean elimination of these materials. Yields are maintained through greater emphasis on cultural practices, IPM, and utilisation of on-farm resources and management. Although the term "low-input farming" has often been used to describe any system of alternative agriculture, it can be seen that it is distinctly different from organic farming, etc. Nevertheless, any system that reduces purchased chemical inputs can be called low-input farming.

As research funded through U.S.D.A.'s LISA/SARE programme has emerged, it is apparent that many Extension programmes are now offering low-input practices as a regular option for growers. Examples of low-input agriculture Extension Service programmes in the United States:

— University of Arkansas reduced herbicide programme for soybeans
— University of Massachusetts low-spray apple orchard programme
— Pennsylvania State University living mulches for vegetables programme

In Oklahoma, speakers have reported at Horticulture Industries Show meetings that:

— poultry litter can replace nitrogen fertilisers in the production of watermelons;

— legume cover crops can supply the total nitrogen requirements of pecan trees; and,
— two timely applications of a synthetic insecticide can produce a full crop of worm-free apples.

In Arkansas, speakers have reported at the Arkansas Society for Horticultural Science meetings that:

— compost amended potting mixes produce superior vegetable transplants than traditional soilless mixes;
— no-till vegetable systems are feasible using reduced herbicide rates to kill cover crops; and,
— subterranean clover living mulches supply nitrogen and weed control in peach orchards.

Integrated pest management is probably the oldest and most widely recognised Extension Service programme devoted to low-input agriculture. However, only recently have the "non-chemical" approaches—such as cultural, mechanical, and biological—within the IPM framework been emphasized over the chemical component. Some programmes, in fact, are now termed "biologically-intensive IPM." In Oklahoma, low-input sustainable agriculture is being practiced on many farms and ranches. Extension efforts are needed to photograph and document these practices for wider distribution. Research efforts are needed to validate practices through on-farm research for wider application.

REFERENCES

American Society of Agronomy, *Decision reached on sustainable ag.*, Agronomy News, January, pg 15, Madison, Wisconsin, 1989.

Bartholomew, G.A., *The roles of physiology and behavior in the maintenance of homeostasis in the desert environment*, Homeostasis and Feedback Mechanisms, Symposia of the Society for Experimental Biology 18:7-29, 1964.

Cox, P.G., N.D. MacLeod, and A.D. Shulman, *Putting sustainability into practice in agricultural research for development*, Submitted to the United Kingdom Systems Society 5th International Conference on

Systems for Sustainability: People, Organizations and Environments, 7-11 July, De Montfort University and The Open University, Milton Keynes, 1997.

Diebel, P., R. Llewelyn, and J. Williams, *An Economic Analysis of Conventional and Alternative Cropping Systems for Northeast Kansas*, Report of Progress 687, Manhattan: Kansas Agricultural Experiment Station, Kansas State University, 1993.

Dobbs, T., M. Leddy, and J. Smolik, "Factors influencing the economic potential for alternative farming systems: case analysis in South Dakota," *American Journal of Alternative Agriculture* 3(1): 26-34, 1988.

Flora, C.B., "Multi-criteria decision-making," In Steiner, J.L., A.J. Franzluebers, C.B. Flora, and R.J. Janke (Eds) *Interactions: Investigating Ecosystem Dynamics at a Watershed Level*. 13-16 April, Proceedings, Athens, Georgia: Georgia Chapter of the Soil and Water Conservation Society, 1997.

2

Environmental Impacts of Mineral Fertilisers

Fertilizers are compounds given to plants to promote growth; they are usually applied either through the soil, for uptake by plant roots, or by foliar feeding, for uptake through leaves. Fertilizers can be organic, or inorganic. They can be naturally occurring compounds such as peat or mineral deposits, or manufactured through natural processes (such as composting) or chemical processes. These chemical compounds leave lawns, gardens, and soils looking beautiful as they are given different essential nutrients that encourage plant growth.

They typically provide, in varying proportions, the three major plant nutrients, the secondary plant nutrients and sometimes trace elements (or micronutrients) with a role in plant or animal nutrition: boron, chlorine, manganese, iron, zinc, copper, molybdenum and (in some countries) selenium. Both organic and inorganic fertilizers were called "manures" derived from the French expression for manual tillage, but this term is now mostly restricted to organic manure. Though nitrogen is plentiful in the earth's atmosphere, relatively few plants engage in nitrogen fixation. Most plants thus require nitrogen compounds to be present in the soil in which they grow.

The use of fertilisers as a regular farming practice began in most European countries in the mid to late nineteenth century but the greatest increase in consumption in these countries occurred in the three decades following World War II. Their increasing use in the developing countries started in the 1960s. In 1960, 87% of the world fertiliser consumption was accounted for by the developed countries, including the USSR and the countries of Central Europe. From 1980 to 1990 consumption tended to stabilise in these regions, apart from the USSR, where it increased until 1988. Population growth had levelled off, almost everyone was adequately fed, world agricultural exports had stagnated due to economic problems in the importing countries and on well managed farms the economic optimum of the available varieties had been reached.

Between 1989 and 1994 fertiliser consumption in developed countries as a whole fell from some 84 Mt nutrient in 1988 to 52 Mt nutrient in 1994. The fall was greatest, by 80% in total, in the formerly communist countries of Central Europe and the former Soviet Union (FSU). Crop production in this region also fell, although not to the same extent. This was because under the centrally planned system fertilisers were used inefficiently and plant available reserves of some nutrients had accumulated in the soil and could now be mined to help feed crops.

In developing countries until the 1960s fertilisers were applied mostly to plantation crops such as tea, coffee, oil-palm, tobacco and rubber, while application to field crops was either small or nonexistent. Even where fertilisers were applied, application rates had to be small in view of the traditional tall cereal varieties which were cultivated at that time. The introduction of highyielding, fertiliser-responsive dwarf varieties in the mid to late sixties gave a considerable boost to fertiliser consumption applied to annual field crops. Unfortunately this development has still not occurred in many countries of sub-Saharan Africa, for economic and climatic reasons and also for lack of suitable varieties.

Since 1960, fertiliser consumption in the developing countries has increased more or less continuously, and today accounts for about 60% of the world total, compared with 12% in 1960, a trend which is continuing. With their rapidly increasing population, many developing countries are compelled to give agricultural production and the development of fertiliser use a high priority. Between 1993/94 and 1997/98 world total fertiliser nutrient consumption increased from 120 to 136 Mt, an average rate of increase of about 3% p.a. Consumption in China, South Asia and Latin America increased by 10, 5 and 2 Mt respectively. But in most countries of sub-Saharan Africa the quantity of fertiliser used is not only very low, but also most of what is used is applied in the commercial, plantation sector. Rates of fertiliser use on food crops are particularly low.

MANUFACTURE OF FERTILISERS

Increasingly the manufacture of fertilisers is a global industry, located near the source of the raw materials or in developing countries with expanding markets for the products. Fertiliser production is decidedly not a monopoly of the developed world.

Nitrogen Fertilisers

The energy required for nitrogen fertiliser production is found around the planet and there is production in every region of the world. However, there has been a trend towards increased production not only in locations where cheap natural gas is available, such as the Middle East and the Caribbean, but also in the main consuming regions, such as South Asia and China.

Phosphate

The main producers of phosphate rock and phosphate fertilisers are the USA, the FSU, China, Africa and the Middle East. Several of these countries are developing countries and the phosphate industry makes an important contribution to their economies. Over the past two decades there has been distinct trend towards the processing of phosphate rock in countries with substantial natural resources of this material, especially in North Africa and the USA, but also in the Middle East and South and West Africa. There have been several plant closures in West Europe, where phosphoric acid production capacity and output have fallen by 60% since 1980, for economic and environmental reasons, particularly the problem of gypsum disposal.

Potash

Potash is produced in the few countries where the ores are located. In 1996 Russia and Belarus accounted for 23% of the world's production, Canada for 35%, West Europe for 23% and Israel and Jordan for 11%, these few regions thus accounting for a total of 92% of world production.

What would happen if mineral fertilisers were not used? The immediate effect of terminating the use of mineral fertilisers is that crop yields would fall to levels sustainable by the soil alone and the relatively small net inputs through organic materials, and that the yields would fall progressively as the soil nutrient reserves are used up, declining eventually to the low levels observed in very long-term trials. In the absence of fertilisers it is likely that cropping systems and management methods would change, but despite all the efforts, it is inevitable that the present structure and output of agriculture could not be maintained. There would simply be insufficient crop nutrients in the overall system. The richer countries may possibly get by but not the poorer countries, and perhaps not the poor in richer countries.

Schmitz and Hartmann established models to make quantitative estimates of the effect of reducing the use of agro-chemicals, including nitrogen, in Germany. They calculated that halving the fertiliser nitrogen application would lead to a 22% reduction of yields in the short-term, 25 to 30% in the medium term, farm profits reduced by about 40%, farm income by 12%, total cereal production reduced by 10%, with a substantial impact on employment in agriculture and the food processing industries, reduced agricultural exports, increased imports, and an increase of the world price of cereals of 5%. For limited reductions in nitrogen use, some ecological benefits would be obtained rapidly but, with across-the-board extensification, the gains would fall and even turn into losses, with a reduction in woodlands and wetlands as these were brought into cultivation. If this is the position in Germany, what should it be in less industrialised countries?

In France in 1850 the average wheat yield was 1000 kg/ha. By 1950 it had reached 1600 kg/ha with a fertiliser consumption of 1.1 Mt total nutrient. By 1973 the average yield was almost 4500 kg/ha, fertiliser consumption 5.8 Mt nutrient, of which 1.8 Mt was N. The average yield between 1994 and 1996 was 6772 kg/ha with the consumption of 4.8 Mt nutrient, of which 2.4 Mt was N. In France there has been a close correlation between the production of cereals and nitrogen deliveries. The effect was made possible by the use of a combination of all the means of production, species and varieties with a high genetic potential, grown on well prepared soil, protected against pests and diseases. Annual yield variations were minimised and production costs reduced. At 1950 yields it is estimated that a household would still spend 50% of its income on food compared with 20% today. France is now the second largest world exporter of agricultural and derived products.

It is sometime salutary, when advocating the termination of some technical advance, to look back at the situation before the advance occurred. Price described the situation in France until the nineteenth century. Prosperity or misery, life or death depended on a good harvest. The last major famine in France was in the early 1700s although "crises de subsistence", when cereal prices increased by 50% to 150%, continued to occur until the mid 1800s. The crises of,1788-89 and 1846-7 were particularly notable in terms of their economic, social and political impact, both preceding popular revolt.

In China, using organic matter to maintain the fertility of the land, rice yields were maintained at 700 kg/ha for thousands of years. During the past 40 to 50 years, using a combination of available organic materials and an ever-increasing input of mineral fertilisers, yields have multiplied six-fold, to reach an average, between 1994 and 1996, of 5958 kg/ha.

In general, it is difficult to estimate the contribution of mineral fertilisers to global agricultural production in view of the interaction of the many factors involved in this biological process. An IFA survey covering developed countries carried out in the 1970s indicated that yields would fall rapidly by some 40% to 50% if fertilisers were no longer applied. According to some Chinese data, fertilisers contribute 40% to 50% of the grain yield, 47% of the cotton yield. V. Smil estimated that, globally, 40% of the protein in the human diet is derived from nitrogen fixed by the Haber-Bosch process for the manufacture of ammonia.

In Japan, A. Suzuki reports that surveys made in 1990 at 92 experimental sites showed that the national average yield obtained without nitrogen applied for several years was 70% of the fertilised plots. Yields decreased gradually over the years. In a long-term trial, after 50 years of NPK fertilisation there was no decrease over the years in the fertilised plots. The yield without fertiliser was about 40% of that of the fertilised plot.

Mackenzie and Taureau gave a typical yield response curve of winter wheat to fertiliser nitrogen in the UK. Even at the economic optimum, where the value of the additional unit of nitrogen equals the value of the crop obtained there from, the response was 3 kg grain per kg N. Without nitrogen the yield would have been 4 t/ ha instead of 7 t/ ha at the economic optimum.

From another series of trials in the UK quoted by the same authors, it was estimated that the yield of wheat increased by 24 kg of grain for every 1 kg of N fertilisers up to the stage where the response started to plateau. Based on a wide range of experiments in a large number of countries, the FAO considered that "it is reasonable to assume that 1 kg of fertiliser nutrient ($N+P_2O_5$ $+K_2O$) produces around 10 kg of cereal grains". K.K.M. Nambiar summarises results from long-term trials in India, from which the following is an extract:

Yields (kg/ha)	*No fertilisers*	*NPK*	*NPK + FYM*
Rice	1751	3607	3994
Wheat	994	3342	3545

DEMAND FOR MINERAL FERTILISERS

Future Demand for Agricultural Products

Population

Between today and, say, 2020, the world's population is going to increase, mostly in the developing countries. According to the World Bank's 1994-1995 population projections, the world's population will increase from 5.7 billion people in 1995 to 7 billion in 2020. This includes increases in China from 1.2 to almost 1.5 billion, South Asia from 1.3 to 1.9 billion, Africa from 0.7 to 1.2 billion. The rate of increase is likely to be highest in Africa but in view of the large population base in South Asia and China, there will inevitably be a substantial increase in these regions. IFPRI estimates that developing countries will account for about 85% of the increase in the global demand for cereals and meat between 1995 and 2020.

FAO projects that 680 million people, 12% of the world's population, could be chronically undernourished in 2010, down from 840 million in 1990-92 but still a very substantial number. 70% of these will be in sub-Saharan Africa and South Asia, especially Bangladesh. In Africa and the Near East, the absolute number of hungry people will increase, though the proportion of the population that is undernourished will decline. Many of these people are the rural poor, who lack the buying power to satisfy their nutritional needs, even when the supplies exist. Women and children are most affected. The issue in their case is to develop agricultural systems which will provide them with sustenance and income.

Income growth

According to the International Food Policy Research Institute, between 1993 and 2020, the global demand for cereals is expected to increase by 41%. The developing countries' demand for cereals for feeding to livestock is expected to double, while demand for direct human consumption is expected to rise by 47%, although the largest absolute increase will be greater for the latter. There will be similar large increases in the demand for other crops.

Global income growth is projected to average 2.7% per year between 1993 and 2020, the growth rate in developing countries being almost twice that of developed countries. Economic growth, rising incomes and urbanisation, particularly in Asia and Latin America, are leading to rapid changes in diets, in favour of more grain-intensive food such as meat, particularly red meat. This leads to a substantial increase in the demand for grain to feed livestock, the impact on cereal requirements being magnified by the rather low feed conversion efficiency of livestock.

IFPRI estimates that the world's farmers will have to produce 40% more grain in 2020, compared with 1995. However, the expansion of the cereal area is unlikely to be more than 5%, almost two thirds of which will be in the difficult region of sub-Saharan Africa. Inevitably most of the higher production must come from higher yields per unit area, which will require a correspondingly larger quantity of plant nutrients, from one source or another.

Fertilisers and food

The exact contribution of mineral fertilisers to agricultural production is debatable but in any case the millions of field trials which have been carried out throughout the world demonstrate clearly their great influence on crop yields. In an article in "The Observer", New Delhi, April 17, 1997, Dr. Swaminathan, a leading Indian scientist, is reported as saying that Fertiliser is the key to securing the food need of more than 1.3 billion Indians by the year 2025. No country has been able to increase agricultural productivity without expanding the use of chemical fertilisers.

Working on a conservative population forecast of 1.3 billion by 2025, India would need 30 to 35 million tonnes of NPK from chemical fertilisers in addition to 10 million tonnes from organic and biofertiliser sources, to produce the minimum food grain need of 300 million tonnes. Scientists have found that there was growing evidence of the increasing deficiency of phosphate and potash in soils, aggravated by the disproportionate application of higher doses of N in relation to P and K. Sulphur has been identified as crucial for optimising the yield from oilseeds, pulses, legumes and high-yielding cereals.

N. E. Borlaug, a Nobel Prize winner, speaking of Africa, stated: "*My 53 years of* experience in low-income developing countries tells me that small-scale farmers are loath to adopt such "low-input, low-output" technologies since they tend to perpetuate human drudgery and the risk of hunger. This certainly has been our experience in Sasakawa- Global 2000, where farmers have overwhelmingly told us they want access to yield- increasing, drudgery-reducing technology, and have proven that they are able and enthusiastically willing to modernise their production".

Several institutions, among them the FAO, IFPRI, UNDP, the US Department of Agriculture and the World Bank, have made projections concerning food security. They differ according to the assumptions made, but essentially they are in agreement that the world supply of food will have to keep growing, and growing rapidly. Agricultural investment, especially in research and advisory services, will be essential if the objective is to be achieved.

ECONOMIC GROWTH

There is today a wide agreement that a necessary condition for economic growth in most developing countries is a productive agriculture—there are some exceptions, but they are few. This was not always the case. In the 1950s the emphasis in development policy was on urban industrial growth, with the agricultural sector being regarded as a source of inputs, mainly labour, for the manufacturing sector. It was only in the 1960s that the positive role of agriculture as an engine of development became accepted. Subsequent events in the 1970s and 1980s reinforced the need for more attention to be paid to agricultural development policies. But even today, some developing countries still do not attach sufficient importance to agricultural development. If agriculture is to be productive, it is evident that the crops should receive, from one source or another, the nutrients they require.

A June 1996 IFPRI study concerning Latin America confirmed how agricultural growth helps the whole economy. When agricultural producers' incomes rise, they spend money on non-agricultural items, creating jobs for others throughout the whole economy. The study found that for every US$ 1 increase in agricultural output in developing countries, the overall economy grows by US$ 2.3. Apart from being good for the national economy, productive agriculture helps to alleviate rural poverty.

Most of the world's poor are rural-based and, even when they are not engaged in their own agricultural activities, they rely on non-farm employment and income that depend directly or indirectly on agriculture. The rural poor make up more than 75% of the poor in many sub-Saharan and Asian countries. Economic growth is strongly linked to poverty reduction. Poverty is itself a form of pollution and, in addition, the poor are often forced to overuse or misuse the natural resource base, in order to meet their basic needs.

Another, February 1994, IFPRI report described the results of a study in seven Asian countries, with widely diverse production environments and agrarian and policy structures, of the effect of technological change in favourable rice-growing areas, on the income of people in unfavourable ones—those by-passed by the new technology. The contributors found that when indirect effects of labour, land and product market adjustments are taken into account, differential adoption of high yielding varieties, HYVs, across environments does not significantly worsen income distribution. As HYV adoption increased the demand for labour in the favourable areas, more inter-regional migration from unfavourable areas took place, which mitigated potentially negative effects by equalising regional wages. Shifts to alternative crops or non-farm employment in the unfavourable areas also contributed to equity.

A 1997 report by the Indian National Council of Applied Economic Research states that India could virtually eliminate urban poverty in a decade if it could sustain an annual economic growth averaging 6.4%. But the report also foresees growing disparities between Indian cities and the countryside where 74% of people live. Agricultural growth is stagnant. The report suggests that the urban 26% of the population will increase to 30% in 2007 but this does not take account of possible accelerated urbanisation prompted by the rising income disparities.

SOIL FERTILITY

As stated by A.E. Johnston soil fertility depends on complex, and often little understood, interactions between the biological, chemical and physical properties of soil. Understanding and quantifying interactions between the biological, chemical and physical properties of soil will become ever more important. He observes that it will be necessary in future to recognise more clearly that there is a distinction between the agricultural productivity and the fertility of a soil:

- Provided soil fertility is at a satisfactory level, within climatic constraints agricultural productivity may be controlled by annual inputs like N and chemicals to control weeds, pests and diseases.
- But soil fertility is frequently controlled by factors which are often difficult to manipulate in the short term, for example, chemical properties like soil acidity and plant nutrient status.

Wherever possible it will be necessary to define critical measures of soil fertility, and then ensure that soils are kept just above them. Below the critical value, loss of yield is a serious financial threat to the sustainability of any husbandry system. Maintaining soils much above the critical value is an unnecessary financial cost to the farmer and may have environmental implications.

Nutrient Depletion

"The loss of soil fertility in many developing countries poses an immediate threat to food production and could result in a catastrophe no less serious than from other forms of environmental degradation". "Agricultural soils lose their fertility by plant nutrient depletion and, in some cases, plant nutrient exhaustion.....a real and immediate threat to food security and to the lives and livelihoods of millions of people. The loss of fertility reduces yields and affects water holding capacity, leading to greater vulnerability to drought."

A fertile and productive soil is the fundamental resource for the farmer and the entire ecosystem. The farmer's objective is to maintain the productivity of his soil. This implies the need for good stewardship on his part; that is, maintaining a good physical structure, organic matter content, good aeration, an adequate moisture content, proper pH and an optimal nutrient status. Management of such a system is complex. The sequence of crops grown, the number of livestock carried on the farm and the cultivation techniques employed by the farmer can either reduce or improve soil productivity.

As regards the plant nutrients, a crop's overall demand and the amount removed from the soil must be replaced, not necessarily annually, but certainly within the overall crop rotation, if soil fertility levels and long-term productivity (sustainability) are to be maintained.

The following paragraph is quoted from an IFPRI report on The World Food Situation published in October 1997.

> "Past and current failures to replenish soil nutrients in many countries must be rectified through the balanced and efficient use of sources of plant nutrients and through improved soil management practices. While some of the plant nutrient requirements can be met through the application of organic materials available on the farm or in the community, such materials are insufficient to replenish the plant nutrients removed from the soils. It is critical that fertiliser use be expanded in those countries where a large share of the population is food insecure. One of the largest environmental problems in Africa today is the gradual decline in the fertility of much of the soil".

The mining of nutrients is part of the cost of producing crops, often a hidden cost which is not passed on to the consumer. Under such circumstances, the use of public funds to help replace the mined nutrients may be justified, especially in situations where the farmer's financial situation is precarious.

Impact of Fertilisers on Soil Structure

It is sometimes claimed that the use of mineral fertilisers has an adverse effect on soil structure. Evidence from very long-term experiments indicates that this is not the case. The aggregating action from enhanced root proliferation and greater amount of decaying residues from well fertilised crops makes soils more friable, easier to cultivate and more receptive to water. S.W. Buol and M.L. Stokes state "*Organic carbon* contents that become lower under inadequate fertilisation appear to recover when adequate fertiliser is applied.

Adequate fertilisation also contributes to greater biomass production tending to protect soil from erosion and providing greater quantities of residue critical to soil aggregation. We therefore conclude that long-term, high-input agriculture has a strong positive effect in improving agronomic properties of soils". Field plots at the Rothamsted Experimental Station in the United Kingdom, which have received chemical fertilisers since 1843, are more productive today than at any time in the recorded past. At the Askov experimental station in Denmark, after 90 years, the plots receiving NPK fertilisers had an 11% higher organic C content than the control plots.

The increase in organic matter content induced by NPK applications resulted in a decrease in soil bulk density and a concomitant increase in total porosity. They conclude that "*The long-term* positive effect of continual application of fertiliser materials on soil organic matter content and soil physical conditions is an important, although often neglected, factor that needs to be considered when *contemplating sustainability"*. In Japan after 50 years of NPK fertilisation there was no decrease over the years in the fertilised plots. The yield without fertiliser was about 40% of that of the fertilised plot.

Soil Acidification

Most nitrogen fertilisers, especially ammonium sulphate and to a lesser extent ammonium nitrate, acidify soils, although some soils are naturally able to cope. The use of organic residues at normal levels of application may not avoid acidification but may slow the process. The acidifying effects of nitrogen fertiliser can be corrected if lime is economically available and is applied. In temperate regions, lime, applied in quantities of tonnes per ha, but less frequently than fertilisers, provides optimum conditions for growth of many temperate crop species.

Apart from neutralising soil acidity, liming improves the availability of other nutrients such as phosphate, and lowers the toxicity of aluminium and manganese. In a long-term experiment in India, with the continuous application of fertilisers without lime, yields fell to zero. When the soil pH was kept near to the optimum, the system became sustainable. In the humid tropics, the lime requirements are high and the effect may not last long due to leaching losses. However, increases in crop yields can sometimes be achieved with minimal applications of lime due to alleviation of aluminium toxicity and/or calcium deficiency and care must be taken to avoid over-liming. In many developing countries agricultural lime is not available at an economic cost. A possible solution lies in the development of cultivars which are tolerant of the direct and indirect effects of soil acidity.

Soil Erosion

Soil erosion is a world-wide phenomenon, but is more

serious in some regions than in others. In regions where a dry season alternates with a season with torrential rains soil erosion can be very severe. At the end of the dry season the soil is likely to have sparse vegetative cover, particularly if the land has been over-grazed by livestock. Under semi-arid conditions wind erosion and desertification are serious problems.

A fertile soil supporting a swiftly growing crop is much less erosion prone than a poor soil supporting a poor crop. The better developed the surface canopy the greater the protection against wind and water action. Because of their vigorous root systems and large residues, high yielding crops help to anchor the soil. The roots and residues returned to the soil enhance productivity by building up organic matter, by improving aeration and by increasing water infiltration rates.

The residual effects of the greater organic matter production are significant also in improving soil aggregation. Land use management appropriate to the topography and rainfall, together with the appropriate use of fertilisers, can make an important contribution to soil conservation. Reduced tillage cultivation practices significantly reduce erosion; the proportion of crop land subject to no-till techniques is expanding rapidly in the USA and certain other countries, for example in Brazil.

TOXIC SUBSTANCES

Phosphate fertilisers often contain small amounts of elements which occur naturally in phosphate rock and are carried through, in the manufacturing process, to the finished product. When the final product is a relatively high-value material destined for use, for example, in the food industry the potentially harmful elements are removed, but to date economic processes for removing these elements economically in fertiliser production have not been found. Among these elements, most attention has been paid to cadmium (Cd).

There is evidence that Cd is slowly building up in some soils. This is of concern because Cd is not essential to plants or animals, and at high levels can be toxic. Sources include atmospheric deposition from industrial processes, sewage sludges, animal manures and phosphate fertilisers. In many European countries about 50% of total Cd input to agricultural soils is from airborne sources. Sewage sludges contain amounts of Cd which can vary from a few p.p.m. to several thousand p.p.m. The use in the manufacture of phosphate fertilisers of phosphate rock with a low cadmium content is one solution, but the total supply is limited. This places the emphasis on developing effective and viable cadmium removal processes and research with this objective continues.

Ultimately the solution could be a combination of the removal of Cd in the manufacturing process and farm management strategies which minimise its potential entrance into the food chain. The uptake of Cd by plants can in fact be affected by many factors, such as soil pH, moisture content, variety etc., which can be controlled by the farmer. There is no immediate urgency because, apart from a few sites heavily polluted by industry, soil cadmium levels are generally well below critical levels. However, the existence of a medium and long-term problem is recognised by the fertiliser industry and studies and research on the subject continue.

WATERS

There is concern that fertilisers are polluting both surface waters and water in aquifers, although the direct impact of the application of mineral fertilisers on the nitrate content of waters is poorly defined. According to Union des Industries de la Fertilisation (UNIFA), 1997, in France, it is estimated that nitrogen fertilisers account for 25% of total mineral nitrogen introduced annually into the ecosystem, or 2.3 Mt N out of a total of 9.4 Mt N. Other major inputs are from nitrogen fixed by leguminous plant (3 Mt N) and animal wastes (2 Mt N).

In a major catchment area in France, 42% of the nitrogen in the water was of agricultural origin (arable and livestock), 49% domestic and 9% industrial. Labeled 15N experiments indicate that not more than 5% of fertiliser nitrogen is lost to water during the growing season, two thirds of it due to incorrect fertilisation practices. In general the extent of losses is not linked directly to recent fertiliser applications. Of the agricultural losses 50% was from soils which were left bare in winter and 33% due to poor cropping practices i.e. losses which could be avoided.

In the mid-1980s the World Health Organisation (WHO) recommended a limit of 50 mg of nitrate, NO_3, per litre of drinking water. They reviewed the recommendation in April 1997 and concluded that, on the basis of the latest scientific evidence, the value of 50 mg per litre should be maintained. The European Union (EU) issued a drinking water directive in 1975. In 1980 another directive set a level of 50 mg per litre.

Then, in December 1991, the EU adopted a directive, known as the Nitrates Directive, concerning the protection of waters against pollution caused by nitrates from agricultural sources. This directive recognised that, whilst the use of nitrogencontaining fertilisers and manure is necessary for EU agriculture, any over-use of fertilisers and manure constitutes an environmental risk. It emphasizes that common action is needed to control the problem arising from intensive livestock production, and that agricultural policy must take greater account of environmental protection.

The objectives of the directive are to ensure that the nitrate concentration in freshwater and groundwater supplies does not exceed the limit of 50 mg NO_3 per litre and to control the incidence of eutrophication. Having set the overall targets, the directive requires individual Member States, within prescribed limits, to draw up their own plans for meeting them. These plans involve the preparation of a voluntary Code of Good Agricultural Practice, the designation of zones vulnerable to water pollution from nitrogen compounds and the implementation of action programmes designed to prevent pollution within those zones.

The measures include a maximum limit for the addition of livestock manure – the main culprit—equivalent to 170 kg nitrogen (N) per hectare. In addition, the periods in which it is acceptable to apply animal manure are defined. The agricultural techniques for keeping nitrate out of water supplies are known. The European Fertiliser Manufacturers' Association (EFMA), for example, has explained these techniques in a code of best agricultural practice.

Surface Waters

The over-enrichment of surface waters leading to an excessive multiplication of algae and other undesirable aquatic plant species, with various undesirable consequences, is a phenomenon known as eutrophication. Whereas phosphate tends to be the limiting nutrient in inland waters, nitrogen tends to be the limiting nutrient in coastal waters.

Coastal waters

In Europe large areas of the North Sea coastlines and areas of the Mediterranean have suffered from eutrophication due to nitrates. In the USA, nitrates and phosphates are suspected of causing *Hypoxia*, or the "Dead Zone" in the Gulf of Mexico. There is a great deal of controversy as to the cause, and even if these nutrients prove to be the cause, they may originate from several different sources apart from mineral fertilisers. Nutrient-enriched water, especially run-off from agriculture, is also incriminated in the *Pfiesteria* problem that killed a large number of fish in Chekaspeake Bay, USA, in the summer of 1997. It is highly unlikely that mineral fertilisers are primarily responsible for either of these problems, but the U.S. fertiliser industry is co-operating fully in the investigations.

Inland waters

Under temperate conditions, in inland fresh water, P is usually the limiting nutrient and very low concentrations can cause problems of eutrophication. Surface water can be enriched with P from both point and diffuse sources. As the amount of P from point sources has declined in recent years, the percentage contribution from diffuse sources has increased. Although it has been generally accepted that most soils fix P strongly, only very small amounts of P have to be lost from soil to maintain the concentration of P in drainage at levels likely to cause environmental problems.

Phosphate in the soil is rather immobile and the loss of water-soluble phosphate through leaching is usually very small (less than 1 kg of P_2O_5 per hectare per year). Ignoring crop removal, the two primary pathways of loss of phosphorus from the soil are by erosion (wind and water) and in run-off.

Under European conditions, the excessive surface applications of animal manures can result in significant losses in particulate matter in run-off. Areas where intensive animal husbandry is practiced may experience the addition of excessive amounts of phosphorus to the soil, usually in the form of heavy applications of animal waste eg. slurry or farmyard manure.

Under these conditions, soils can have such a high content of phosphorus that losses may increase. In tropical lakes, there is evidence that nitrogen can be the limiting nutrient. Phosphate concentrations in the water are often higher than in temperate regions while inputs of N from surrounding land may be low. Surface runoff from cropland, pasture and forest, can contribute to phosphate loading of surface waters. Best management practices are highly effective in eliminating this possibility and, at the same time, allow for the most efficient use of crop fertilisation.

Potash

Unlike nitrogen and phosphate, potash has no known deleterious effect on the quality of natural waters.

ROLE OF AIR

Nitrogen can be lost from agricultural systems in three forms which may cause pollution; nitrate loss by leaching, ammonia volatilisation and nitrous oxide loss during denitrification or nitrification. Ammonia loss to the atmosphere and its subsequent deposition contributes to the eutrophication of natural habitats and marine waters and to the acidification of soils and lakes as the NH_4 is converted to NO_3. Losses by denitrification are harmless if the end product is nitrogen gas but if the resulting gas is nitrous oxide there is a contribution to the greenhouse effect and to depletion of ozone in the upper atmosphere.

Ammonia

H. Kirchmann observed that ammonia deposition from the atmosphere may enrich terrestrial and aquatic ecosystems. On average in West Europe 92% of all ammonia originates from agriculture. About 30% of the nitrogen excreted by farm animals is released to the atmosphere from animal houses, during storage, grazing and application of animal wastes to the soil. Ammonia emissions from growing arable crops are low, but emissions can be higher from decomposing crops. Composting results in high ammonia losses.

Deposition of ammonia takes place over areas where lower amounts would have been supplied. This deposition together with that from nitrogen oxides decreases biodiversity, but it can increase carbon storage in sediments and forest soils. Near very large animal farms, local toxic effects damaging the surrounding vegetation can occur. Ammonia deposition contributes to acidification of soils as ammonia is nitrified to nitrate and then nitrate is lost by leaching.

Ammonia can also react in the atmosphere with sulphur oxides to form ammonium sulphate, which precipitates into the soil with rainfall and causes acidification. Although most emissions of ammonia are from manure or natural sources, experiments demonstrate that nitrogen losses to the atmosphere in the form of ammonia following the application of urea can amount to 20% or more, under temperate conditions. Losses occur when the urea is not incorporated into the soil immediately after spreading and they are particularly high on calcareous soils. The expanding practice of reduced tillage cultivation is increasingly the surface application of urea. Losses are even higher, up to 40% or more, under tropical conditions, on flooded rice and on perennial crops to which the urea is applied on the surface, such as bananas, sugar cane, oil palm and rubber.

Greenhouse Gases

Carbon dioxide (CO_2), methane (CH_4) and nitrous oxide (N_2O) are the three most important greenhouse gases. They absorb solar radiation rather than allowing the heat to be radiated away from the earth. Their impact as greenhouse gases, or global warming potential (GWP), is a function of two factors, their "radiative forcing" and on their lifetime in the air. Taking the GWP of CO_2 as 1, that of CH_4 is 21 and that of N_2O is 310.

Carbon dioxide

Fixation of carbon dioxide by photosynthesis is the source of organic carbon in crops and eventually in soils. Crop production practices that enhance photosynthetic activity improve the retention of carbon. Decomposition of organic matter releases carbon dioxide back to the atmosphere. Good fertilisation and tillage management practices improve the net gain of carbon to the soil. Recent estimates indicate that agricultural and forested land in the Northern Hemisphere is now a net sink for carbon dioxide from the soil/plant complex due to increased vegetative growth. According to E. Solberg, for every pound of nitrogen applied as fertiliser, 10 to 12 pounds of carbon can be sequestered. The rapidly increasing use of reduced tillage systems is helping to rebuild soil organic matter, hence increasing the quantity of carbon stored.

Nitrous oxide

Nitrous oxide has a greenhouse effect and is considered to be detrimental to the ozone layer. According to experts of the Intergovernmental Panel on Climate Change (IPCC), nitrous oxide is responsible for 7.5% of the calculated greenhouse effect caused by human activity. Its concentration in the atmosphere is increasing at a rate of about 0.2% per annum. Soils are the major global source of N_2O accounting for some 65% of all emissions; they are the result of microbial processes. Nitrogen fertilisers can be a direct and indirect source of N_2O emissions; IPCC currently assume an N_2O emission factor of 1.25% of fertiliser N applied, but with a nine-fold range, from 0.25% to 2.25%. In general, fertiliser management strategies that increase the efficiency of N uptake by crops are likely to reduce emission of N_2O to the atmosphere.

Methane (CH_4)

Methane production mainly stems from wetlands, paddy

fields, gastric fermentation within ruminant animals, animal wastes, domestic sewage and abiotic sources such as coal mining and natural gases etc. The direct impact of chemical fertilisers on methane emission is not clear. In the USA it is estimated that agricultural sources account for 29% of US methane emissions. Of the agricultural emissions ruminant animals account for 62%, animal waste for 32% and rice paddies for 5%.

There are indications that cultivation and N fertilisation decreases the rate at which CH_4 is taken up from the atmosphere by soils, thus contributing towards atmospheric methane levels, but the amounts involved are small relative to total sources. Suzuki reports that rice paddy fields in Japan are thought to account for about 10% to 30% of total methane emissions from all sources. In rice paddies methane is formed by the anaerobic decomposition of organic matter. The addition of readily decomposable organic matter significantly increases methane emissions.

NUTRIENT LOSSES AND EFFICIENCY

In view of the large quantities involved, inefficiencies in fertiliser use represent a substantial economic loss. For example, given that about 80 Mt of N were used in world agriculture 1996, a 20% loss with a wholesale price of US$ 0.66 per kg of N in urea, amounts to US$ 10.6 billion. Excessive losses of nitrogen and phosphate to waters and of nitrogen to the atmosphere also present an environmental risk. Plants acquire most of their nutrients either from soil reserves or from recently added fertilisers or organic manures. Assessing whether added nutrients are used efficiently is usually done by the difference method.

$$\% \text{ efficiency} = (\text{recovery})$$

$$\frac{\text{Au in crop given Aa} - \text{Au in crop without A}}{\text{A}} \times 100$$

where A is the nutrient tested at amount Aa, and Au is the

amount of A in the crop grown with and without A. Calculated in this way the recovery of added nutrients is very dependent on the yield of the crop receiving the nutrient being tested, and the amount of nutrient which is supplied by the soil. Different time scales can be used. Usually only one crop or year is considered but, for soils in which reserves of plant available nutrients can be accumulated, it is appropriate to consider a longer time span.

A. Finck considered that the proportions of fertiliser nutrients taken up by the crop during the season of application are as follows:

— Nitrogen: 50-70%
— Phosphate: 15%
— Potash: 50-60%

Data from Rothamsted Experimental Station in the UK for phosphorus, show that from 1843 to until the 1970s, the overall P offtake was about a third of that applied, similar to Finck's data. But with today's yield of 8.5 t/ha grain, the P offtake in grain plus straw is almost equal to the annual application, of 35 kg P/ha, although not necessarily from the P applied for that particular crop. Similarly the increased yield of winter wheat now removes in grain plus straw most of the K applied each year (90 kg K/ha).

Efficiency of Nitrogen

When assessing the efficiency of nitrogen, it is important to take account of the fact that the plant is, in fact, in competition with the soil microbial population. This is especially so in soils in which organic matter is accumulating. Pilbeam collated data from experiments in which 15N labeled fertiliser, with N in different forms, was applied at different growth stages to rain-fed crops of wheat grown in different locations worldwide. The proportions of nitrogen taken up by the crop and soil respectively varied widely in response to differences in rainfall and evaporation between the locations but the proportion of applied labeled N that was unaccounted for, and presumably lost from the crop-soil system, was largely independent of variations in climate. The unexplained loss of fertiliser N ranged from 10% to 30%, average 20%.

A.E. Johnston reported that 15N experiments at the Rothamsted Experimental Station in the U.K. showed that, on average, about 20% of the applied N had been incorporated into soil organic matter between application and harvest. These two factors, the unavoidable and partly unexplained N loss averaging 20% and the average of 20% incorporated into the soil, correspond to Finck's 50 to 70% estimate of plant uptake. The nitrogen incorporated into the soil as organic matter may subsequently be mineralised and become available to subsequent crops. And because it is not easy to predict both the amount and time at which this organic nitrogen is mineralised it is difficult to give accurate recommendations for fertiliser nitrogen applications.

Although 50% to 70% of the applied nitrogen can be taken up under the controlled conditions of experimental stations, in practice, losses can be much greater. R. S. Paroda stated, in relation to India, that "The nitrogen use efficiency varies from 20 to 25% in rice, 21 to 45% in maize, 45 to 50% in wheat. A 1% increase in the recovery rate of N fertiliser would save 98 Kt N, equivalent to 1 Mt food-grains. In the case of phosphate, recovery varies *from 15% to 20%"*. In an earlier paper, R.S. Paroda et al. observed that in the ricewheat systems of Asia, nitrogen fertiliser efficiency is estimated at around 30 to 40%. For micronutrients, such as zinc, the efficiency seldom exceeds 10 to 15%.

The following text is extracted from a paper by Peoples et al. (1995).

> *"Unfortunately, fertilizer sources are not utilized* efficiently in agricultural systems, and plant uptake *of fertilizer N seldom exceeds 50% of the N applied*. One of the principal reasons for

> the poor efficiency in fertiliser use is that a proportion of the N applied (up to 89%) is lost from the plant-soil system. Fertiliser N can be lost by leaching, erosion and run-off, or by gaseous emissions. The relative importance of these processes can vary widely, depending on the agricultural system and the environment. Similarly the relative importance of NH_3 volatilisation and denitrification varies considerably and depends on the agro-ecosystem, form of fertiliser N used, crop management imposed and the prevailing environmental conditions. It is puzzling that farmers in so many countries tolerate the poor efficiencies of fertiliser use. They persist with poor agronomy when simple management practices are available which could increase the efficiency of N uptake and decrease costs of production. Special problems arise with crops such as rice, cotton and sugar cane which receive large applications of N but which also lose large amounts of N by denitrification and NH_3 volatilisation. When the economic situation is good, farmers are unconcerned about applying excess amounts of N, but the environmental consequences of this wasteful practice certainly need to be considered....Many approaches are now available to control the losses of fertiliser N by NH_3 volatilisation and denitrification".

In work in China, in 25 on-farm experiments, the average plant recovery of N by an early rice crop averaged 29% (range 10% to 65%), compared with 41% in a experimental station trial. In the case of farmers' practice 5 kg grain per kg of N applied was obtained compared with 24 kg on the experimental station. On a late rice crop, recovery averaged 5%, range 0% to 12%. The author estimates that only 60% of the potential yield is achieved in intensive rice-growing areas of Asia, with very high N losses to the environment. In trials on rice in Vietnam, average recovery efficiency of applied N was 40% in farmers' practice, but with a yield of only 11 kg grain per kg N applied. At another site, with improved agricultural practices a recovery of 69% was obtained by farmers, and 30 kg grain per kg N.

Much can be achieved by improving management practices. Matson et al., working on wheat in an intensive agricultural region of Mexico, found that an improved management system reduced gaseous loss of N from about 14 kg N/ha to virtually zero. A greater plant uptake can also be achieved with new varieties. A. Suzuki reported that a high yielding variety of rice in Japan took up about 160 kg N per ha whereas a commonly grown variety took up 130 kg/ha.

Phosphate and Potash

Until a few years ago, it was believed that phosphate (and potash) "fixed" by the soil went over into plant-unavailable, inert and hence useless forms. There has, however, been a change of perception in recent years. Experiments have shown that in many soils reserves of plant available P and K can be built up over time. Soils enriched with these reserves frequently gave larger yields than soils without the reserves. Hence the plant-uptake figure of 15% for P underestimates the long-term efficiency of phosphate fertilisers. Phosphate (and potash) residues accumulated in the soil are not necessarily lost—but this is not a reason for accumulating these residues unnecessarily.

There are critical values of phosphate and potash below which yield decreases appreciably and which represent a financial loss to the farmer. To accumulate P and K in the soil above these critical levels is an unnecessary cost for the farmer. It may also pose an environmental risk, in that soil lost by water or wind erosion to streams, rivers and lakes takes its nutrient load with it. Further work is required to establish these critical soil levels under different conditions. Work to improve the plant uptake of applied P and K is also required. Phosphate has both direct and indirect effects.

The increased availability of phosphate has a positive effect on the quantity and quality of agricultural outputs. Through indirect interaction effects, phosphate increases the response of agricultural production to the other inputs such as nitrogen and potassium and has positive effects on biological nitrogen fixation, soil organic matter maintenance, water-holding capacity, soil erosion control and other soil physical and chemical properties. All of these positive effects result in increased agricultural output, sustained productivity and land conservation.

Product Differentiation

In fact, improvements in fertiliser use efficiency can be detected in most agriculturally advanced regions, but this can be attributed to improvements in cultivation practices, techniques of fertiliser application and crop varieties. Apart from some developments in coated, controlled release fertilisers and nitrification inhibitors, there has been little significant change in the fundamental nature of the main fertiliser products for many years, or even decades. There is little incentive to invest in the research and development of a bulky, low-priced commodity which offers little scope for product differentiation.

Slow release fertilisers

Controlled release nitrogen fertilisers have agronomic advantages, especially in tropical regions, and in regions with light-textured soils and under heavy rainfall or irrigation, where N losses are particularly large. However, to date the cost of slow-release fertilisers has limited their use to high value crops such as vegetables. The use of controlled release nitrogen fertilisers on field crops is most advanced in Japan, on rice, whose production is heavily subsidised.

The amount of fertiliser required by rice is commonly applied in 3 to 4 applications. This is labour intensive and in order to reduce the number of applications, coated slow-release fertiliser have been studied. Experimental results so far have indicated that a single application of the whole amount of coated fertilisers in a basal application gives yields comparable to those with traditional split applications. Also the efficiency of nitrogen use could be improved by 10% to 20% due mostly to an increase of about 16% in the plant uptake of N.

Nitrification and urease inhibitors

Environmental restrictions in certain developed countries may encourage more farmers to use nitrogen or urease inhibitors in association with nitrogen fertilisers in areas where it is essential to reduce losses for environmental reasons. The application of urea (or a urea/ammonium nitrate solution, UAN) amended with a urease inhibitor would generally permit a substantial reduction in nitrogen losses to the atmosphere, and consequently also in the application rates, without affecting growth and yield of fertilised crops.

The future, and in particular the wider use of slow or controlled release fertilisers, and nitrification and urease inhibitors primarily depends on the development of new, effective, low-price and non-toxic products. Even if promising new products were developed, due to lengthy, time-consuming tests and data collection for registration purposes, the introduction to the market would take several years. It should also be taken into account that giving advice to farmers on how to use them correctly would be very expensive.

Nutrient absorption enhancers

Increasing the absorption of applied nutrients by the plants, as opposed to the soil, is a means of increasing fertiliser use efficiency. J.L. Anders and L.S. Murphy presented work with a polymer which shows improved nutrient uptake and nutrient use efficiency.

Bio-fertilisers; microbial inoculants

Microbial seed inoculants, commonly but wrongly called bio-fertilisers, are able to enhance biological nitrogen fixation or to solubilise soil phosphate. The inocultants are claimed to be cost effective, eco-friendly and renewable, and generally capable of supplementing chemical fertilisers in sustainable agricultural systems. It has long been known that the inoculation of efficient strains of the symbiotic *Rhizobium* can be beneficial for leguminous pulses and oilseeds.

Free-living organisms such as *Azobacter* and *Azospirillum* have proved effective for rice and certain other crops. The problem with inoculants is that establishment and therefore effectiveness depend on the natural conditions and on the skill of the user. As regards the product itself, the inoculant is a living material and there are problems due to the need to select the most effective strains, the difficulty of quality control, the short shelf-life, the need to avoid high temperatures in storage etc.

As regards phosphate, several phosphatesolubilising bacteria are known to mobilise significant quantities of soil phosphate that would otherwise not be available to the plant, but their effectiveness is variable and not predictable. Vesicular-arbuscular mycorrhizae have favourable effects on P uptake but much more research and development is required before reliable commercial products can be made available. At present it is difficult to produce mycorrhizae in bulk.

The Food and Fertiliser Technology Center for the Asia and Pacific Region report that, while there is an increasing interest in Asia in the use of N-fixing and P-solubilising bacteria, the technology of producing and using them is still at an early stage. While some are very effective, farmers often find themselves paying large sums for useless products. There is a very large number of different microorganisms in microbial products and they are often not identified, whereas some are crop-specific. They tend to be heavily promoted and there is a great need for standards and for simple and accurate ways of measuring their effectiveness.

Efficient Use of Fertilisers

Efficient fertilisation is important from both an economic and an environmental point of view. It is synonymous with minimising nutrient losses to the environment, while optimising crop yields. Excess nitrogen not taken up by the crop is likely to be lost to the environment. The quantities and relative proportions of the different nutrients required by particular crop and soil conditions must be respected. The challenge is to maintain the fertility of soils despite the increasing demands placed on them. Fertiliser use efficiency has in fact been improving in the developed countries.

Fertigation

A technique which enables growers to maximise the use of water resources and to increase the efficiency of fertiliser use is "fertigation". This technique is particularly appropriate for high value crops under arid and semi-arid conditions; it is widely used in Israel. It involves the addition of soluble fertilisers into irrigation systems, preferably using a "drip system" which allows uniform water distribution and feeding of the crop. The fertiliser can be applied to the crop whenever it is needed. The initial investment cost may be expensive, but all irrigation systems are expensive. Maintenance of the system and its management requires skilled labour.

Balanced Fertilisation

If any plant nutrient, whether a major nutrient or a micro-nutrient, is deficient crop growth is likely to be affected. One definition of balanced fertilisation is "the nutrient mix which gives the optimum economic return". This may be at high levels in intensive agriculture, or at comparatively low levels in less favourable circumstances. In either case balanced fertilisation is necessary for use efficiency. The application of nitrogen fertilisers tends to be preferred by farmers, because of their relatively low cost per unit of nutrient, their widespread availability, and the quick and evident response of the plant. However, the increased yields deplete the soils of the other plant nutrients removed by the harvested crops, unless they are replenished.

Research at IRRI in the Philippines has shown that while the application of an adequate quantity of N increased the yield of rice paddy 2.9 times, it also resulted in the removal of 2.6 times more P, 3.7 times more K and 4.6 times more S from the soil, compared to the amounts removed from unfertilised soil. In due course, these nutrients have to be replaced if the yields are not to suffer. The same applies to micronutrients. In a 1995 FAO document *"Rice and the* environment: production impact, economic costs and *policy implications"* it is stated that incorrect fertiliser use in much of Asia, unbalanced in favour of nitrogen, results in lodging, greater weed competition and pest attacks, with a financial loss varying from 4% to 30% of the rice price.

"Fertiliser use has been increasing rapidly in Pakistan over many years but there is a stagnation crop production. This seems to be due largely to the incorrect use of fertilisers. Farmers have been applying high amounts of nitrogen, but only small quantities of phosphate. Other fertilisers, such as potash and micronutrients are hardly used at all. Organic sources are not being properly integrated with mineral fertilisers. Under such conditions, the soil is depleted and it takes more nitrogen every *season to obtain the same crop"*.

Site Specific Fertiliser Application

The need for rational and sustainable land use, especially in regions subject to severe population pressures, emphasizes the need for effective land use planning. The classification of land types according to their agricultural suitability, together with the implementation of soil conservation measures, was used with great success to combat the erosion and desertification problems encountered in US agriculture in the 1930s. Fertiliser recommendations should take into consideration specific agro-climatic and environmental conditions.

General recommendations need to be adjusted to the conditions of the particular field. They depend on factors such as soil characteristics, cultivation practices, quality and quantity of irrigation water, ground water table, crop rotations and the managerial capacity of the farmer. The expected yield level of a crop is an important consideration. The rapid progress in information technology during the past decade, including Geographic Information Systems (GIS) and computerised mapping, offers the possibility of agro-ecological zoning which can help in a preliminary selection of crops and technologies, including appropriate fertiliser use, suited to the local conditions and the problems encountered.

In highly developed systems, precision farming may use satellite communication and detailed field and crop information to improve farm operations and nutrient efficiency by means of the site-specific application of fertilisers. Soil analysis and crop deficiency diagnosis to facilitate the fine-tuning of fertiliser rates to actual crop requirements are of fundamental importance to precision agriculture. Precision agriculture does not, of course, necessarily require sophisticated machinery and satellite-positioning systems. Farmers in developing countries could well improve the precision of their plant nutrient programmes given soil testing facilities and sound advice.

INTEGRATED SYSTEMS

R.N. Prasad. stated *"The ultimate goals or* the ends of

sustainable agriculture are to develop farming systems that are productive and profitable, conserve the natural resource base, protect the environment and enhance health and safety in the long run. Traditional agricultural systems that met the test of sustainability in the past have not been able to respond adequately to today's growth in demand for agricultural commodities required by the current population pressures of humankind and animals and rapidly declining resources of good quality arable land and water resources.

The basic principles of soil management for sustainable agricultural systems are:

— Replenish nutrients removed
— Maintain the physical condition
— No build-up of weeds, pests and diseases
— No increase in soil acidity or toxic elements
— Soil erosion must be controlled to be equal or less than the rate of soil genesis".

"Integrated farming" or in French "Agriculture raisonnee" is the combination of farming practices—including the use of rotations, cultivation, choice of variety and skilful, precise use of fertilisers and crop protection products—with measures to preserve and protect the environment. The best combination must be specific to each farm. The concept is based on the German model (FIP) founded in 1987. In the UK LEAF (Linking Environment and Farming) has support from the government, farming groups, research organisations, the fertiliser industry, environmental campaign groups and consumer organisations.

In France, FARRE, the "Forum de l' agriculture raisonnee respectueuse de l'environnement" is gaining widespread support. "Integrated farming" takes systematic and simultaneous account of the environmental aspects, the quality of the produce and the profitability of the farm. Its aim is to develop an agriculture which is sustainable but which corresponds to the farmers' needs and to society's expectations. Soil and water resources and biodiversity are respected. Fertilisation and crop protection techniques which minimise adverse environmental impacts are adopted. Animal health and well-being, management of effluents and waste, optimal use of water resources and erosion control are all taken into account.

"Alternative farming" has ideological undertones whereas "integrated farming" aims to make optimum use of the best known techniques. A. Leake, of the CWS Farms Group in the UK, has stated "Integrated farming is a recent development but is already showing promise in its ability to deliver high yielding crops, cost effectively with reduced environmental impact. Such a system offers a real alternative for European agriculture compared to conventional high input systems and organic low output farming".

Land Planning

There are examples in certain regions of the EU and elsewhere, particularly in water catchment areas, of successful co-operation between farmers, water authorities and agricultural advisory services which have enabled local environmental targets to be achieved. "Landcare" in Australia is an example of a successful land management programme. There are comparable programmes in certain other countries, such as Brazil, India, South Africa....

Ferti-Mieux

In France, in 1980, the French government established COMIFER, the "Comite français d'etude et de development de la fertilisation raisonnee", whose objective is to promote rational fertilisation "fertilisation raisonnee", making use of all scientific, technical and practical means. It is composed of representatives of public authorities and educational establishments, farmers' organisations, fertiliser producers and distributors.

"Ferti-Mieux" defines the steps to be taken in order to obtain, in a given catchment area, a progressive change in agricultural practices which would minimise the risk of polluting water. Participation is voluntary. It is based on a consensus, on a national basis, between the Ministry of Agriculture, the farmers' associations, the fertiliser associations, other Ministries etc., and at the local level, between farmers, advisors, consumers, water agencies etc. An approved Ferti-Mieux operation, which respects the guidelines, is recognised by a label, which is attributed, for one or two years, by three different national bodies. The label provides a guarantee, for farmers, advisors, financial bodies and the general public.

Integrated Plant Nutrition Systems, IPNS

In 1996 IFA published a document prepared by R. Dudal entitled "Plant Nutrients for Food Security", drawing attention to the importance of the effective management of plant nutrients as a major component of agricultural development. A substantial part of the document is concerned with integrated plant nutrition systems (IPNS) and related subjects. He defined Integrated Plant Nutrition as "an approach which adapts plant nutrition to a specific farming system and particular yield targets, the physical resource base, the available plant nutrient sources and the socioeconomic background".

The sources of plant nutrients may be mineral fertilisers and/or biological nitrogen fixation and/ or organic materials, depending on the circumstances. Recommendations of an FAO-IFFCO (Indian Farmers Fertiliser Cooperative) International Seminar on IPNS for Sustainable Development held in New Delhi, India, in November 1997 were as follows:

— The development of IPNS requires an improved service to the farmers, in the form of technical advice, inputs, credit, marketing facilities, public investment in agriculture. IPNS should:
— address both increased productivity and increased profitability for farmers, with special attention paid to the alleviation of poverty in rural areas,
— integrate the maintenance of natural resources and rehabilitation of these resources where needed and the enhancement of productivity in agriculture,
— be system-oriented and should in particular take account of the interactions between plant nutrient supply and water supply, between plant nutrient supply and the control of pests and diseases,
— improve the availability of energy for the rural population, in order to save fuel wood and organic materials used as an energy source, be science based, associating agronomy, ecology and social science,
— use a "Farming Systems" approach, not limited only to cropping systems.

Leguminous Plants as a Source of N

Leguminous crops provide a substantial input of nitrogen into the ecosystem. It has been estimated that biological nitrogen fixation supplies about 30 to 40 Mt nitrogen per year, which compares with about 80 Mt from fertiliser nitrogen. The contribution of leguminous plants to crop nutrition has long been known and made use of in traditional systems. Leguminous crops have a high requirement of phosphorus and potassium and this has to be supplied. The microorganisms, living in symbiosis, receive their energy from the plant in return for the nitrogen they produce. They are not efficient converters and the energy used in the natural fixation process is at the expense of the yield of the crop i.e. yields of leguminous crops tend to be low.

The possibility of using leguminous crops as a source of nitrogen is of particular interest to small-scale farmers who cannot afford to purchase nitrogen fertilisers. They are not, however, cost-free. If the production of the leguminous crop is not otherwise economic, they occupy land which might be put to better use. M.E. Summer draws attention to trials in Africa in which a rotation comprising two or three years of *Sesbania* fallow followed by maize gave a spectacular increase in maize production compared with unfertilised, continuous maize. However, he points out that it is still necessary to apply phosphate. Nor is nitrogen derived from leguminous crops more environmentally friendly than that provided by mineral fertilisers; in fact it is likely to be less friendly.

The amount, rate and timing of nutrient release is difficult to control. In trials on grass/clover leys in the UK, following the ploughing of the leys, winter wheat was grown and soils were sampled during the winter and spring. It was calculated that between 110 and 250 kg N per ha were leached as the ley length increased from 1 to 6 years. In the average, through drainage on this soil, the amount of N raised the nitrate concentration in the drainage from just below 200 mg/l to 400 mg/l, eight times the EU limit for nitrate in drinking water.

Green manures, particularly of nitrogen-fixing leguminous plants, are an important source of nitrogen. However, they can be unattractive from the farmer's point of view if they do not produce salable or comestible product. Farmers with only a small area of land can scarcely afford to use part of it unproductively. *Green manures* are labour intensive. They provide significant amounts of nitrogen but require the application of phosphate and other nutrients. They are no more environmentally friendly than mineral fertilisers; for example, there is evidence that nitrous oxide emitted from fields after legumes in amounts similar to those from fertilised crops. The release of the nitrogen fixed by leguminous crops is difficult to control.

Azolla, a floating aquatic fern associated with nitrogen-fixing blue-green algae is used as a source of nitrogen for flooded rice. Used as a green manure the optimum application amounts to several tonnes per hectare. The fern requires a considerable quantity of water and of phosphate and it cannot withstand high temperatures. Green manures, such as *Sesbania* have long been used in China but since the date of that tour the use of nitrogen fertilisers there has increased from 8 to 23 Mt N.

NUTRIENT ACCOUNTING

Using comparisons of nutrient inputs and outputs as a yardstick for environmental correctness of fertilising practices in scientific publications and studies started during the 1980s. Different types of nutrient balances came into use. The most common is a comparison of nutrient inputs and outputs at farm gate level. The account examines the relationship between applied nutrients and nutrients removal in the harvested crop. It considers all nutrients, whether of mineral or organic origin, which may be applied.

The system ideally should also consider changes in soil nutrient levels and, in some cases, admissible losses. Nutrient accounting is being developed by the OECD as one of the environmental indicators. These are national indicators K. Parris and L. Reille require careful interpretation. The nutrient balance indicator needs to be used in conjunction with indicators on farm nutrient management, soil quality and water quality.

Livestock wastes contain substantial amounts of plant nutrients. It is therefore evident that all sources of nutrients should be into account when determining rates of mineral fertilisation. Accounting systems based on the nutrient input and output are used in some European countries as a measure for the environmental performance of farms, particularly in countries with a manure disposal problem.

In Denmark, since 1994, farmers have had to prepare fertilisation plans and the amount of nitrogen that may be applied to each type of crop is regulated. Another requirement is that 65% of the cultivated area must be covered by a green crop in winter. There are heavy fines in the event of infringement. In Germany a federal "Regulation of fertiliser use" model regulation came into effect in January 1996. The model must now be implemented in the individual Federal States. In Norway fertiliser plans are now compulsory. In the Netherlands a compulsory nutrient accounting scheme is starting in 1998. Nutrient applications over the maximum will be taxed.

Nutrient accounts may indicate a deficit as well as a surplus. The exercise would therefore be useful also in developing countries where soils are being mined of their nutrients. A substantial proportion of the nutrients which find their way into the manure produced by intensive livestock units originate from animal feed which has been imported from other regions of the world, thus depleting the nutrients in the soils of the exporting regions. But in many developing countries soils are also being exhausted just to provide a subsistence for their cultivators.

HEALTH

Human Health

Nitrates

Normally drinking water provides from 10 to 30% of the nitrate ingested, the remainder coming from vegetables, fruit and meat. The proportion of nitrate converted to nitrite also varies widely, but about 7% is typical. Nitrate in drinking water is considered to be a public health problem because nitrate rapidly reduces to nitrite in the body. The nitrite oxidises the blood haemoglobin which is unable to transport oxygen to the tissues; this can manifest itself in babies up to six months old, causing the blue-baby syndrome. It is normally due to the conversion by microbes of nitrates into nitrites in the feeding bottles as a result of defective hygiene. The occurrence of the syndrome is now rare, although, for some reason, cases still occur in Hungary and Romania.

Another concern is that nitrite may react with compounds in the stomach to produce N-nitroso compounds, particularly nitrosamines, which have tested positive in animal carcinogenicity experiments. Ingested nitrate is absorbed rather quickly from the upper gastrointestinal tract and most is subsequently eliminated in the urine. About 25% of the nitrate in blood is secreted by salivary glands, and the microbial flora of the oral cavity reduce some of it to nitrite.

It is never possible to prove zero risk, but it should be taken into account that doses of nitrosamines which have proved carcinogenic in animals far exceed those to which humans are exposed. No link to cancer in humans has been demonstrated. In fact, there is now evidence that some nitrate is beneficial. Many pathogens are susceptible to nitrite under acid conditions. Phosphate and potash have no known adverse effect on human health. Both elements are an important constituent of living organisms. Far from potash intake having any harmful effect, it may have a beneficial effect on human health in reducing hypertension.

Product quality

Some claim that crops grown with "artificial fertilisers" have less taste and are less healthy than, for example, organically grown produce. In fact, the plant does not distinguish between the original source, mineral or organic, of its nutrients they are all taken up in the same chemical form. However, according to a literature review by K.

Woese et al., conventionally produced or fertilised vegetables may have a higher nitrate content compared with organically produced or unfertilised vegetables, especially those green and root vegetables which are nitrophilic. Vegetables of organic origin also tend to have a higher content of dry matter. Regarding all other parameters which determine nutritional values and sensory assessment, they noted that differences between the two systems were not significant or that the results were contradictory so that conclusions could not be drawn.

Over-fertilisation should evidently be avoided but correct fertilisation can have a positive impact on the quality of agricultural produce. For example, the mineral, protein and vitamin contents of crops may be improved if judicious fertilisation corrects a previously existing inadequate level of nutrient availability. The baking quality of wheat, the brewing quality of barley and the colour, crispness and textural character of various vegetable crops benefit from appropriate fertilisation, as can the calibre, taste and flavour of fruits. Potassium improves the tuber quality of potatoes and the sugar content sugarcane. Sulphur increases the protein content in grain and the oil content of oil-seed crops. And so on.

Plant Health

Fertiliser use at excessive rates has deleterious effects on crop growth. Examples are the lodging of cereals and the low sugar content of sugar beet resulting from excess quantities of nitrogen, nutritional disorders involving trace elements such as zinc due to excessive phosphate fertiliser and lime; impeded seed germination and seedling injury from too much soluble fertiliser salt adjacent to the seed row; the acidifying action of nitrogen fertiliser on soil, and increased incidence of plant and pest attacks with excessive nitrogen fertiliser. If the nitrogen application leads to acidification of the soil, it can induce aluminum and manganese toxicity if compensating lime is not applied.

As regards crop diseases, the most important impact of nitrogen is on vigour and plant growth. These two factors have an important impact on plant susceptibility to many diseases. Vigorous plants with rapid growth are generally more sensitive to obligate parasites and some pathogens are specifically more aggressive towards vigorous plants. However, most of the necrotic pathogens attack less vigorous plants with nitrogen deficiency. Balanced fertilisation provides excellent protection. The time of application of fertilisers is important. A wrong timing may induce substantial growth of foliar parts of plants and maintain high humidity conditions in the crop canopy which are favourable to disease development. Phosphorus application seems to favour plant protection against diseases, either by correcting a deficiency in soil phosphorus, and thereby inducing a better growth of the plant, or by speeding up the maturation process, disfavouring some pathogens like downy mildew that effects the young tissues.

Potash can increase the efficiency of use of other nutrients by plants, particularly of N. Potash has a beneficial effect on the quality of a wide range of crops, especially in terms of improved protein quantity and quality. Potash can decrease the incidence of plant diseases and reduce abiotic stresses, particularly cold stress. The element may have a direct action on pathogen penetration, lesion size and on inoculum density. An indirect effect of potassium on disease development is to stimulate the healing process, to increase the resistance to cold, and also to delay the maturity and senescence of fruits. There is no known pollution or health hazard due to the use of potash fertilisers in agriculture. However, the application of potassium chloride to chloride-sensitive crops should be avoided, as should its use on certain saline soils.

Calcium may have an effect on the cell wall of plants by making them more resistant to pathogen penetration. A deficiency in calcium increases the sensitivity of plants to many fungi.

BIODIVERSITY

Plant and animal communities may be directly affected by changes in their environment through variations in the quality of water, air and soil and sediments and through disturbance by noise, extraneous light and changes in vegetation cover. Such changes may directly affect the biosphere, for example habitat, food and nutrient supplies, breeding areas, migration routes, vulnerability to predators, or changes in herbivore grazing patterns, which may then have a secondary effect on predators. Soil disturbance and removal of vegetation and secondary effects such as erosion and siltation directly affect communities, and also lead to indirect effects by upsetting nutrient balances and microbial activity in the soil.

A common long-term effect is loss of habitat, which affects both faunal and floral communities, and induces changes in species composition and primary production cycles. For example, in some countries, population pressure is leading to the cultivation of unsuitable, fragile soils. Tropical forests, growing on soils that are usually highly weathered, are being felled on this account. A large proportion of the Amazon forest, for example, grows on poor soils, which deteriorate further and rapidly after deforestation. There is ample scope for improving agricultural productivity on more suitable land elsewhere in Brazil, thus avoiding the opening of new areas of the Amazon forest and even allowing some degraded areas to revert to natural forest.

In Indonesia, land settlement schemes have involved the felling of rain forest, following which soils have deteriorated rapidly. With adequate fertilisation and good management practices, it has been shown that this land can be rehabilitated, thus avoiding the necessity of clearing additional rain forest and preventing further soil erosion and degradation.

Over-grazing is one of the major causes of soil erosion and the grazing livestock population is tending to increase. The increased production of fodder, with appropriate fertilisation practices, is an excellent means of reducing the pressure of livestock on grazing land. An indiscriminate reduction in fertiliser use would require farmers to use more crop acres to maintain, or increase, present production levels. This would require the use of less productive, more fragile soils.

Urbanisation increases carbon emissions, whereas plants absorb carbon. Mannion noted, however, that, with intensification, the agricultural area is tending to decrease in much of the developed world, with corresponding increases for example in the forested area. This represents a net increase in the carbon sink. But in the developing world, the agricultural area is tending to increase, tropical forest is being transformed into agricultural land, and agricultural land is being lost to urbanisation. This development clearly reduces the vegetative carbon sink, as well as resulting in a loss of biodiversity and genetic resources.

Currently progress is being made in many regions of the world in implementing diversity friendly agricultural practices in soil conservation, withdrawing production from marginal areas, mastering chemical and nutrient runoff, breeding crop varieties which are genetically resistant to diseases, pests and abiotic stresses.

In the USA, the 1996 Farm Act created new programmes such as the Environmental Quality Incentives Programme, the Wildlife Habitat Incentives Programme and the Farmland Protection Programme. A number of other policy options intended to promote sustainability are in various stages of adoption. In 1996 the Agri-environmental "Accompanying Measures" of the EU accounted for over 2 billion ECU, or about US$ 1.8 billion.

Until recently, the biology of what happens in the root zone—the rhizosphere—was relatively neglected. M.J. Swift writes *"Soil management has* been dominated by what may be termed an 'environmental management' paradigm. Crop production is seen as being regulated by its physicochemical environment which can be altered and managed by physical means and the introduction of inorganic chemicals to suit the crop's needs. In recent years an alternative concept of 'biological management' has been emerging which focuses on the manipulation of biological population and processes in soil as well as on its physico-chemical properties. At no location on the earth's surface has it been possible to assess the full biological diversity of the community of soil organisms.....The conventional approach to agricultural management seeks to bypass or even inhibit these biological regulators and often disrupts or destroys ecosystem stability and resilience. A biologically-driven approach provides a broader, ecological concept of soil management which is more readily translated across scales from plot to ecosystem and landscape. It is not only distinct from the green revolution physico-chemical paradigm but also from that of organic agriculture in that it does not eschew petrochemically derived inputs but rather focuses on the efficiency of their use. Ecosystem science provides a framework which integrates the functional attributes of biological population with their physical and chemical environments".

It is known that the heavy use of nitrogen fertiliser inhibits the activity of symbiotic nitrogen fixing organisms such as *Rhizobium* species. If the legume plant is well supplied with nitrogen from the soil and/or mineral fertiliser, it is a less efficient nitrogen fixer; many legumes do not nodulate in the presence of a high soil nitrate level

It has also been contended that fertiliser use, particularly nitrogen application, may inhibit the soil microorganisms from mineralising available soil organic matter.

Soil invertebrates perform an important function in the maintenance of soil fertility. Mineral fertilisers have been accused of having an adverse effect on the earth worm population. It is certainly possible to demonstrate the lethal effects of fertiliser salts and anhydrous ammonia when applied in contact with a living worm. But only a small portion of the soil habitat occupied by worms is in direct contact with applied fertilisers, and consequently, the proportion of the total population detrimentally affected is small. A possible adverse impact on the earthworm population could result from the acidification of soils through the application of certain nitrogen fertilisers not balanced by liming; earthworms are inhibited by soil acidity. However, some researchers have established that the greater supply of fresh organic material obtained through fertilisation is of far greater significance to the earthworm. Size and numbers of earthworms invariably increase as soils are brought from a low to high level of fertility through effective fertilisation.

The circumstantial evidence of the experiments in which mineral fertilisers have been applied continuously for a very long period of time, in a fully sustainable system, a priori indicates that correct fertilisation practices do not harm soil flora and fauna essential for crop production.

REFERENCES

Addiscott, T.M., "Fertilizers and nitrate leaching", In Hester, *R.E. and Harrison R.M. (eds.) Agricultural Chemicals and the Environment,* The Royal Society of Chemistry, Cambridge, U.K. pp. 1-26, 1996.

Buol S.W. and Stokes M.L, "Soil profile alteration under long-term, high-input agriculture", in *Replenishing soil fertility in Africa* Soil Science Society of America, Special Publication N° 51, pp. 97 to 109, Madison, Wisconsin, USA, 1997, 1997.

3

Soil Fertility and Sustainable Agriculture

ESSENTIAL NUTRIENTS

Plant growth is the result of a complex process whereby the plant synthesizes solar energy, carbon dioxide, water, and nutrients from the soil. In all, between 21 and 24 elements are necessary for plant growth. The primary nutrients for plant growth are nitrogen, phosphorus, and potassium (known collectively as NPK). When insufficient, these primary nutrients are most often responsible for limiting crop growth. Nitrogen, the most intensively used element, is available in virtually unlimited quantities in the atmosphere and is continually recycled among plants, soil, water, and air. However, it is often unavailable in the correct form for proper absorption and synthesis by the plant.

In addition to the primary nutrients, less intensively used secondary nutrients (sulphur, calcium, and magnesium) are necessary as well. A number of micronutrients such as chlorine, iron, manganese, sinc, copper, boron, and molybdenum also influence plant growth. These micronutrients are required in small amounts (ranging from a few grams to a few hundred grams per hectare) for the proper functioning of plant metabolism. The absolute or relative absence of any of these nutrients can hamper plant growth; alternatively, too high a concentration can be toxic to the plant or to humans.

CHARACTERISTICS OF SOIL

The capacity of soils to be productive depends on more than just plant nutrients. The physical, biological, and chemical characteristics of a soil—for example its organic matter content, acidity, texture, depth, and water-retention capacity—all influence fertility. Because these attributes differ among soils, soils differ in their quality. Some soils, because of their texture or depth, for example, are inherently productive because they can store and make available large amounts of water and nutrients to plants. Conversely, other soils have such poor nutrient and organic matter content that they are virtually infertile.

The way soils are managed can improve or degrade the natural quality of soils. Mismanagement has led to the degradation of millions of acres of land through erosion, compaction, salinisation, acidification, and pollution by heavy metals. The process of reversing soil degradation is expensive and time consuming; some heavily degraded soils may not be recoverable. On the other hand, good management can limit physical losses. Good management includes use of cover crops and soil conservation measures; addition of organic matter to the soil; and judicious use of chemical fertlisers, pesticides, and farm machinery. Organic matter content is important for the proper management of soil fertility. Organic matter in soil helps plants grow by improving water-holding capacity and drought-resistance. Moreover, organic matter permits better aeration, enhances the absorption and release of nutrients, and makes the soil less susceptible to leaching and erosion.

PLANT NEEDS

Plants need a given quantity and mix of nutrients to flourish. The higher the yield, the greater the nutrient requirement. A shortage of one or more nutrients can inhibit or stunt plant growth. But excess nutrients, especially those provided by inorganic fertlisers, can be wasteful, costly, and, in some instances, harmful to the environment. Effective and efficient management of the soil storehouse by the farmer is thus essential for maintaining soil fertility and sustaining high yields. To achieve healthy growth and optimal yield levels, nutrients must be available not only in the correct quantity and proportion, but in a usable form and at the right time. For the farmer, an economic optimum may differ from a physical optimum, depending on the added cost of inputs and the value of benefits derived from any increased output.

NUTRIENT CYCLE

Soil nutrient availability changes over time. The continuous recycling of nutrients into and out of the soil is known as the nutrient cycle. The cycle involves complex biological and chemical interactions, some of which are not yet fully understood. A simplified version of this cycle of plant growth, based on Smaling, is shown in Figure 1.

The simplified cycle has two parts: "inputs" that add plant nutrients to the soil and "outputs" that export them from the soil largely in the form of agricultural products. Important input sources include inorganic fertlisers; organic fertlisers such as manure, plant residues, and cover crops; nitrogen generated by leguminous plants; and atmospheric nitrogen deposition. Nutrients are exported from the field through harvested crops and crop residues, as well as through leaching, atmospheric volatilisation, and erosion.

The difference between the volume of inputs and outputs constitutes the nutrient balance. Positive nutrient balances in the soils could indicate that farming systems are inefficient and, in the extreme, that they may be polluting the environment. Negative balances could well indicate that soils are being mined and that farming systems are unsustainable over the long term.

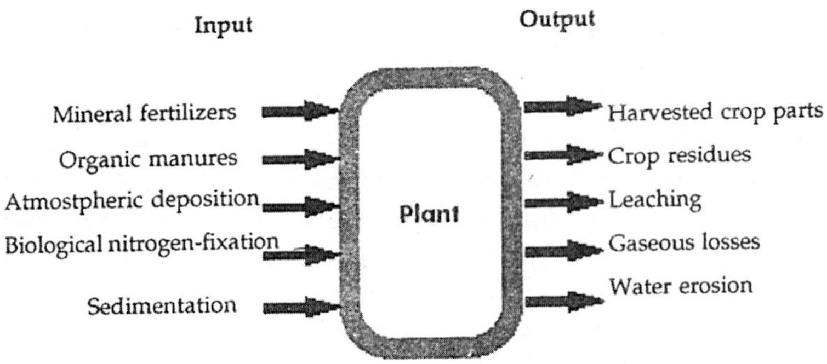

Figure 1: The plant-nutrient balance system

In the latter instance, nutrients have to be replenished to maintain agricultural output and soil fertility into the future. The inexpensive supply of nutrients in the form of inorganic fertlisers was a key factor, along with improved modern seed varieties and adequate supplies of water, in the substantial increase in yields that exemplified the Green Revolution of the 1960s and 1970s.

PLANT NUTRIENT USE AND SOIL FERTILITY

Despite the continued development of new and improved modern varieties and greater use of chemical fertlisers, yield growth began to slow in the latter part of the 20th century. The world's annual cereal yield growth rate has declined from an average of 2.2 percent in the 1970s to 1.1 percent in the 1990s (Table 1). Wheat yields in Asia grew at an average annual rate of 4.3 percent during the 1970s. But during 1990–1997, wheat yields dropped to the far slower growth rate of 0.7 percent per year. After rapid growth of almost 2.4 percent per year during the 1980s, Asian rice yield growth fell to 1.5 percent per year in the 1990s.

This global slowdown has raised concerns that yield growth may have reached a plateau or begun to decline in many of the world's most fertile areas. In Sub-Saharan Africa, the situation is even more dramatic, with cereal yield growth decreasing steadily from 1.9 percent during the 1970s to 0.7 percent in the 1990s. These declines in Sub-Saharan Africa are partly attributable to poor soil management, which in turn has been accentuated by a number of other factors, including inappropriate policies, insufficient commitment to investment in agricultural research, falling agricultural prices, demographic pressures, land availability constraints, and ill-defined property rights. The cumulative effect of all these factors has led to increased soil mining.

Slowdown in Investment

During the 1970s, governments and intergovernmental organisations gave investment in agricultural research a relatively high priority. In Asia, for example, real expenditure on public agricultural research and development grew at an average annual rate of 8.7 percent. During the 1980s, however, real expenditure growth in Asia slowed to 6.2 percent per year. Spending in Sub-Saharan Africa slowed even further, from 2.5 percent per year during the 1970s to 0.8 percent during the 1980s. For developed countries, investment growth slowed as well from 2.7 percent per year in the 1970s to 1.7 percent in the 1980s. Without continued investment in research to propel the development of yieldenhancing technologies and sustainable agricultural management practices, crop yield growth could eventually stagnate and soil fertility could degrade to irrecoverable levels.

Table 1— Annual cereal crop yield growth rates, 1970s -1990s

Crop	*Region*	*1970s*	*1980s*	*1990s*
			(percent)	
Wheat	Asia	4.33	3.71	0.72
	Latin America	0.60	3.40	2.36

	Sub-Saharan Africa	3.54	0.92	- 0.81
	World	2.10	2.78	0.42
Rice	Asia	1.61	2.42	1.55
	Latin America	0.70	2.97	3.71
	Sub-Saharan Africa	0.02	2.51	0.56
	World	1.49	2.37	1.54
Maize	Asia	3.43	2.75	1.55
	Latin America	1.49	0.61	3.82
	Sub-Saharan Africa	2.26	1.72	2.09
	World	3.19	0.60	1.76
Cereals	Asia	2.90	2.79	1.46
	Latin America	1.69	1.28	3.12
	Sub-Saharan Africa	1.90	0.56	0.66
	World	2.18	1.79	1.12

Investment in new irrigation infrastructure has also declined from its peak during the late 1970s and early 1980s. Both average annual public expenditures and annual lending and assistance for irrigation systems from international development agencies have fallen. Numerous factors have contributed to the reduction in irrigation investment, including the poor performance of some past investments, fewer low-cost irrigation sites for potential development, increased real capital costs for construction of new irrigation systems, and environmental concerns such as the spread of salinisation.

In Sub-Saharan Africa, lack of water is a serious restriction as well. Currently only 4 percent of cultivated land in Sub-Saharan Africa (5.3 million hectares) is irrigated, of which 70 percent is in Madagascar, Nigeria, and Sudan. Insufficient water retards nutrient availability and plant growth. Although the potential exists to bring an additional 20 million hectares of land under irrigation in Sub-Saharan Africa, technical, financial and socioeconomic constraints have slowed this expansion.

Declining commodity prices during the 1980s also reduced governmental and intergovernmental incentives to make agriculture-related investments. Prices for coffee and cocoa during the 1990-92 period fell to 39 percent of their nominal price in 1980-82. Similarly, the prices of wheat, maize, and rice in 1990-92 declined to 60, 61, and 50 percent of the prices, respectively, in 1980-82.

Overapplication of Nutrient

Concern has also grown in recent years that the use of fertlisers, particularly inorganic fertlisers, can lead to serious environmental consequences. Environmental contamination of this type, however, is largely a problem in the developed world and a few regions of the developing world. As fertlisers make up a small share of the total production costs in many developed countries, farmers often apply fertilizer in excess of recommended levels in order to ensure high yields. Overapplication of inorganic and organic fertlisers is estimated to have boosted nutrient capacity in the soil by about 2,000 kilograms of nitrogen, 700 kilograms of phosphorus, and 1,000 kilograms of potassium per hectare of arable land in Europe and North America during the past 30 years. Such oversupply of nutrients can lead to environmental contamination, which often has negative consequences for humans and animals.

Overapplication of nitrogen, for example, allows the nutrient to be carried away in groundwater and to contaminate surface water and underground aquifers. Ingestion of nitrate can be toxic to humans and animals when it is transformed within the body into nitrite, which affects the oxygen-carrying ability of red blood cells. Evidence also suggests that nitrite and the carcinogenic compounds it can create may also lead to goiter, birth defects, heart disease, and stomach, liver, and esophagus cancers.

Leaching and run-off of surplus nitrogen and phosphorus into rivers, lakes, and inlets can cause eutrophication—an excess accumulation of nutrients in water that promotes the overproduction of algae. Excess surface algae deprive underwater plants of sunlight, which in turn alters the aquatic food cycle.

The decomposition of dead algae by bacteria reduces the amount of oxygen in the water available for fish. Nitrogen also escapes into the atmosphere in the form of nitrogen gas and various nitrous oxides. In the upper atmosphere, nitrous oxides react to form acid rain, which can harm crops, acidify soil and water, and damage property. Cumulative application of acidifying ammonia-based fertlisers, together with acid rain, also contributes to soil acidification.

Evidence is mounting that excessive fertilisation can damage the environment. In the United Kingdom, some 1.6 million people get water with nitrate levels that exceed guidelines, and Danish, Dutch, and German coastal regions show signs of eutrophication. Furthermore, USDA (U.S. Department of Agriculture) estimates that agriculture causes nearly two-thirds of the pollution in U.S. rivers and that runoff from excess plant nutrient application causes close to 60 percent of the pollution in lakes.

SOIL FERTILITY PROBLEMS

While the overapplication of inorganic and organic fertlisers has led to environmental contamination in a number of areas in the developed world, insufficient application of nutrients and poor soil management, along with harsh climatic conditions and other factors, have contributed to the degradation of soils in Sub-Saharan Africa.

Harsh climatic conditions contribute to soil erosion in several parts of Sub-Saharan Africa. Rapid water evaporation and inadequate and highly variable rainfall, for instance, deprive plants of the water necessary for growth. High atmospheric temperatures, strong light, and heat-retentive, sandy soils can combine to make the local environment too hot for proper plant growth.

Powerful, dry wind gusts may also damage plants through both lodging (which causes plants to fall over and die before harvest) and evaporation. Together, these harsh climatic factors, coupled with poor soil management, have reduced soil fertility by contributing to soil and water erosion.

Slight to moderate erosion slowly strips the land of the soil, organic matter, and nutrients necessary for plant growth. This degradation increases the opportunity for drought and further erosion because it reduces the water-infiltration and waterholding capacity of the soil. Severe erosion may create gullies that interfere with farm machinery use. It may also lead to the conversion of land to lower-value uses, or its temporary or permanent abandonment. Off-farm erosion can lead to siltation in watersheds and a decline in water quality.

In such an environment, effective soil, water, pest, and crop management becomes absolutely essential. But economic and other pressures often make it difficult for farmers and their families to efficiently manage the soil for long-term profitability and sustainability.

Insecure and crumbling tenure arrangements also contribute to declining soil fertility. Communal rights to graze land without any effort to maximise long-term returns has led to serious overgrazing, which is reported to be the main cause of human induced degradation in Africa. Ill-defined property rights and insecure tenure rights have also reduced the incentive for farmers to undertake soil fertility-enhancing investments.

Secure tenure arrangements can help induce investment in soil fertility to reap the long-term reward of sustained high crop yields and greater profits. In Niger, for example, secure land for growing millet accounts for 90 percent of manured fields. These fields received an average of 307 kilograms per hectare of manure, while unsecured millet fields received only 186 kilograms per hectare. Sharecropping may contribute to land degradation as well.

In Ghana, for example, sharecroppers have put enormous pressure on soil fertility to realise immediate high yields in order to pay land rents. Farmers in such situations discount the future at very high rates, thereby reducing the incentive for long-term investments in improved soil fertility.

Demographic pressures and land availability constraints have also contributed to the decline in yield growth and soil fertility. With increasing population, the traditional techniques for renewing soil fertility, such as slash-and-burn and long-term fallowing, are not as feasible as they once were. The need for subsistence production and income are such that land can no longer be taken out of production for substantial periods to allow for natural nutrient replenishment. Nor are animal manures and crop residues usually sufficient for replacing lost nutrients.

Other traditional soil fertility management techniques also generally fall short of the nutrient requirements or today's intensive agricultural practices. For example, in order to provide 150 kg of plant nutrients to fertilise one hectare of land, a farmer could apply either 300 kg of inorganic NPK fertilizer, or 20 to 25 metric tons of crop residue grown on 6 to 10 hectares of land, or 18 metric tons of animal manure generated from crop residue grown on 10 to 15 hectares of land. Under normal circumstances, farmers generally do not have the resources to produce sufficient organic fertlisers to replace all the nutrients removed at harvest time. Indeed, it has been estimated that without the improved input technologies developed during the 20th century, the planet would feed no more than 2.6 billion people, less than half its present population.

Cumulative Effect of Negative Nutrient Balances

Cumulative negative nutrient balances heighten the impact of climatic factors, insecure tenure arrangements, and land and demographic pressures on soil fertility. In 1993 7 million metric tons of nitrogen, phosphorus, potassium, magnesium, and calcium were depleted from soils in the low-income countries of Bangladesh, Indonesia, Myanmar, Philippines, Thailand, and Vietnam. In Sub-Saharan Africa net annual nutrient depletion was estimated at 22 kilograms of nitrogen, 2.5 kilograms of phosphorus, and 15 kilograms of potassium per hectare during 1982-84. Estimates in Sub-Saharan Africa indicate a net loss of about 700 kilograms of nitrogen, 100 kilograms of phosphorus, and 450 kilograms of potassium per hectare in about 100 million hectares of cultivated land over the last 30 years.

The cumulative effect of yearly negative nutrient balances on crop yields is often seen through the impact of soil erosion on productivity. In the United States, for example, if present erosion rates continue and inputs are managed effectively, productivity could decrease by 5 to 8 percent over the next 100 years, with regional variations ranging from 0.7 to 7.1 percent or 3 to 10 percent.

Modern integrated management and conservation practices could lower projected erosion-related productivity losses to about 2 percent over the next 100 years. But this is a largely insignificant decrease when considered against annual productivity gains in the U.S. of about 1 percent per year from new technology and improved management. In many parts of the developing world where poor soil conservation and management methods prevail, however, long-term productivity is projected to decline substantially unless soil management practices improve. In Africa and Asia, past erosion has reportedly reduced average yields by 10 to 20 percent over the past 100 years.

In especially fragile areas, such as in southeastern Tunisia, erosion has reduced long-term productivity by more than 50 percent. If erosion at this rate continues unabated, yields may decrease by another 16.5 percent in Asia and 14.5 percent in Sub-Saharan Africa by 2020. Despite the cumulative effect of negative nutrient balances, overall yields in Africa have increased. From 1960 to the mid-1990s, wheat yields more than doubled from 0.7 to 1.8 metric tons per hectare, while maize yields rose from 1.0 to 1.7 metric tons.

Together with the limited adoption of new technologies, the mobility of the Sub-Saharan Africa farmer has been a major factor in the improvement of yields, albeit at the cost of soil degradation. Between 1973 and 1988, arable and cropped land increased by 14 million hectares, forest and woodland area fell by 40 million hectares, and pasture land remained stable. Thus 26 million hectares have been lost to desertification or abandoned.

The effect of reduced soil fertility remains generally hidden because farmers abandon nutrient-depleted land to clear and farm uncultivated, marginal land. Once the land constraint becomes binding, however, as in the case of the Mossi plateau region in Burkina Faso, yields and production decline, thereby also contributing to migration to urban areas.

Effect of Declining Soil Fertility

The effects of declining soil fertility on yield growth are particularly visible in Africa, where the most serious food security challenges exist and lie ahead. The low level of chemical fertilizer use, decline in soil organic matter, and insufficient attention to crop nutrient studies contribute the most to the loss of soil fertility in the region.

In comparison to the rest of the world, fertilizer use in Sub-Saharan Africa is low and declining. In 1996, Sub-Saharan Africa consumed only 1.2 million tons of fertilizer, (equivalent to 8.9 kilograms per hectare of arable land). By comparison, global fertilizer use reached approximately 135 million tons in 1996, equivalent to 97.7 kilograms per hectare. While fertilizer use per hectare in developing countries continued to increase at a rate of 3.1 percent per year during the 1990s, in Sub-Saharan Africa it declined.

The fall in consumption has been most dramatic in West Africa and the developing countries of Southern Africa. Fertilizer use would probably be even lower if foreign aid were not available. More than half of the nitrogenous, phosphate, and potash fertilizer consumed in developing Africa is imported in the form of aid. In 1990, 22 of 40 Sub-Saharan Africa countries received all their fertilizer imports as aid.

High import prices contribute to the low level of fertilizer use in Sub-Saharan Africa. High fertilizer prices arise from small procurement orders (tenders for less than 5,000 metric tons are common), weak bargaining power, and high freight and international marketing costs. Special mixes tailored for African needs, and other micronutrient additions, such as sulfur or boron, may add an additional US$35 per ton or about 20 percent to the price.

When coupled with high transportation costs due to poor infrastructure, the domestic prices of chemical fertilizer are such that one kilogram of nitrogenous fertilizer can cost the typical African farmer between 6 and 11 kilograms of grain, compared with 2 to 3 kilograms of grain in Asia.

Most African farmers practice low-input agriculture that depends on organic matter in the soil to sustain production. Soil organic matter plays an important part in establishing the intrinsic properties of a soil, which make plant growth possible. Soil organic matter helps sustain soil fertility by improving retention of mineral nutrients, increasing the water-holding capacity of soils, and increasing the amount of soil flora and fauna.

Continuous cropping and erosion reduce the level of soil organic matter. Low-input systems can maintain and enhance soil organic matter though crop rotation and intercropping, the application of animal and green manures, fallowing, and reduced tillage. But as pressure on land and crop intensification increase, these options do not remain practical.

The adoption of intercropping and crop rotation techniques is often constrained by the extent of land and technology available and by the lack of knowledge about optimal management techniques. Farmers need to know how to combine organic fertlisers with chemical fertlisers, apply improved pest and weed management techniques, and adopt highyielding crop varieties.

Insufficient attention to effective crop nutrition and soil fertility management studies has also made it difficult to improve yields in Africa, even when improved germplasm has been made available. More research, expanded extension, and greater integration of knowledge could provide farmers with a stronger incentive to improve yields, maintain soil fertility, and sustain agriculture.

Influence of Government

Although agriculture is increasingly recognised as the engine of economic growth in Sub-Saharan Africa, the level of government commitment to it is low. In the past governments often have penalised agriculture through a variety of mechanisms, including export and import taxes, foreign exchange controls, export licensing requirements and controls, and bureaucratic marketing boards. Food subsidies have allowed governments to keep food prices low, often to appease vocal urban constituents, but at the expense of rural producers. Such policies and practices have reduced farmers' incentives to increase local foodgrain production and use modern inputs to improve productivity.

The lack of competition and heavy government regulation, along with structural factors such as inadequate institutional and physical infrastructure and underdeveloped research and extension systems, have often made fertilizer distribution systems inefficient and ineffective in meeting farmers' needs.

Structural adjustment programs (SAPs) have been instituted in many countries partly in response to these and other market failures. SAPs seek to reallocate resource use in order to improve economic efficiency and social welfare. Among other things, the programs have devalued exchange rates, the immediate effect of which has made imports such as fertlisers more expensive, which in turn has often increased farmers' costs markedly.

Nitrogen-to-maize price ratios in Ghana, Tanzania, and Zambia, for example, were substantially higher during the 1990s, after the SAPs were instituted, than during the 1980s, when price controls and subsidies were in effect.

The SAPs and higher input prices consequently have reduced the profitability of using fertilizer to increase the production of foodgrains for domestic consumption. Farmers growing export crops, though, have benefited from the restructuring of currencies and increased their fertilizer use.

But, given the vast acreage devoted to food crops compared with the modest area under export crops, the devaluation of currencies and the reduction of fertilizer subsidies on balance have militated against increased application of imported fertlisers. Regardless of the type of crop produced, and despite the cost of fertilizer, three factors appear to be key in determining whether farmers use fertlisers.

First, fertlisers should help farmers obtain sufficiently high yields. Second, farmers should be near major towns, where agricultural input distributors are located, in order to benefit from lower prices for inputs and higher prices and lower marketing costs for outputs. Third, farmers should be able to store at least part of their output, so that they can take advantage of higher out-of-season prices. When these three factors are in place, farmers are more willing and able to use fertilizer to increase income and sustain soils for the long term.

CHALLENGES AND RESPONSES AT THE FARMER'S LEVEL

Declining soil fertility and mismanagement of plant nutrients have made the task of providing food for the world's population in 2020 and beyond more difficult. The negative consequences of environmental damage, land constraints, population pressure, and institutional deficiencies have been reinforced by a limited understanding of the biological processes necessary to optimise nutrient cycling, minimise use of external inputs, and maximise input use efficiency, particularly in tropical agriculture. But some responses can ameliorate these difficulties.

The responses highlighted here comprise the approach commonly known as integrated nutrient management (INM). The implementation of INM responses will require a concerted and committed effort by actors from a variety of sectors, including the private and public sectors, scientific and policy organisations, and industrialised and developing countries.

Integrated Nutrient Management

Goal of INM

Sustainable agricultural production incorporates the idea

that natural resources should be used to generate increased output and incomes, especially for low-income groups, without depleting the natural resource base. In this context, INM maintains soils as storehouses of plant nutrients that are essential for vegetative growth.

INM's goal is to integrate the use of all natural and man-made sources of plant nutrients, so that crop productivity increases in an efficient and environmentally benign manner, without sacrificing soil productivity of future generations. INM relies on a number of factors, including appropriate nutrient application and conservation and the transfer of knowledge about INM practices to farmers and researchers.

Plant nutrient application

Balanced application of appropriate fertlisers is a major component of INM. Fertlisers need to be applied at the level required for optimal crop growth based on crop requirements and agroclimatic considerations. At the same time, negative externalities should be minimised.

Overapplication of fertlisers, while inexpensive for some farmers in developed countries, induces neither substantially greater crop nutrient uptake nor significantly higher yields. Rather, excessive nutrient applications are economically wasteful and can damage the environment. Underapplication, on the other hand, can retard crop growth and lower yields in the short term, and in the long term jeopardise sustainability through soil mining and erosion.

The wrong kind of nutrient application can be wasteful as well. In Ngados, East Java, for example, the application of more than 1,000 kilograms per hectare of chemical fertilizer could not prevent potato crop yields from declining. Yields on these fields decreased more than 50 percent in comparison with yields on fields where improved soil management techniques were used and green manure was applied.

The correction of nutrient imbalances can have a dramatic effect on yields. In Kenya the application of nitrogenous fertilizer on nitrogen-poor soils increased maize yields from 4.5 to 6.3 tons per hectare, while application of less appropriate phosphate fertlisers increased yields to only 4.7 tons per hectare. Balanced fertilisation should also include secondary nutrients and micronutrients, both of which are often most readily available from organic fertlisers such as animal and green manures.

Lastly, balance is necessary for sustainability over time. Figure 3 shows that wheat yields become uneconomical after 5 years when only N fertilizer is applied. Even annual field applications of NP and NPK fertlisers were insufficient to sustain yields over the long term. Only when both lime and NPK fertilizer were applied did yields increase and fields remain productive despite continuous cultivation. Coupled with other complementary measures, effective nutrient and soil management can help to reclaim degraded lands for long-term use in some cases.

Heavy fertilizer applications on moderately degraded soil can not only replenish nutrients, but can produce about 7 tons per hectare of maize and about 6 tons per hectare of grain straw, which long-term studies in Iowa have shown can increase organic matter content in the soil. Experiments in Ghana and Niger have demonstrated that by increasing the longevity and productivity of suitable agricultural land, the application of inorganic and organic fertilizer reduces the need to cultivate unsustainable and fragile marginal lands.

Nutrient conservation and uptake

Nutrient conservation in the soil is another critical component of INM. Soil conservation technologies prevent the physical loss of soil and nutrients through leaching and erosion and fall into three general categories. First, practices such as terracing, alley cropping, and low-till farming alter the local physical environment of the field and thereby prevent soil and nutrients from being carried away. Second, mulch application, cover crops, intercrop-ping, and biological nitrogen fixation act as physical barriers to wind and water erosion and help to improve soil characteristics and structure. Lastly, organic manures such as animal and green manures also aid soil conservation by improving soil structure and replenishing secondary nutrients and micronutrients.

Improved application and targeting of inorganic and organic fertilizer not only conserves nutrients in the soil, but makes nutrient uptake more efficient. Most crops make inefficient use of nitrogen. Often less than 50 percent of applied nitrogen is found in the harvest crop. In a particular case in Niger, only 20 percent of applied nitrogen remained in the harvest crop. Volatilisation of ammonia into the atmosphere can account for a large share of the lost nitrogen. In flooded rice, for example, volatilisation can cause 20 to 80 percent of nitrogen to be lost from fertilizer sources. These losses can be reduced, however.

Deep placement of fertlisers in soil provides a physical barrier that traps ammonia. The use of inhibitors or urea coatings that slow the conversion of urea to ammonium can reduce the nutrient loss that occurs through leaching runoff, and volatilisation. With innovations of these kinds, better timing, and more concentrated fertlisers, nutrient uptake efficiency can be expected to improve by as much as 30 percent in the developed world and 20 percent in developing countries by the year 2020.

Untapped nutrient sources

If used appropriately, the recycling of organic waste from urban to rural areas is a potential, largely untapped, source of nutrients for farm and crop needs, especially on agricultural lands near urban centres. For example, environmentally undesirable wastewater has been used to irrigate fields and return nutrients and organic matter to the soil. Like organic manure, urban waste sludge is a source of primary nutrients, albeit a relatively poor source in comparison with commercial fertlisers.

Stabilised municipal waste sludge typically contains about 3.3 percent nitrogen, 2.3 percent phosphorus, and 0.3 percent potassium, although some concentrations can reach as high as 10 percent nitrogen and 8 percent phosphorus on a dry weight basis.

Actual nutrient content, however, varies widely and depends on the source of the waste. Urban waste also has a number of other benefits. Like other organic manures, it helps improve soil structure by adding organic matter to the soil. It is also a source of the secondary nutrients and micronutrients that are necessary in small quantities for proper plant growth. In addition, urban waste transforms material that would otherwise be slated for costly disposal into a useful farm product.

Urban waste needs to be treated carefully because it may contain heavy metals, parasites, and other pathogens. The buildup of heavy metal concentrations in the soil can be cause for concern. While trace amounts of some heavy metals play a critical role in plant metabolism, excessive amounts have reduced crop yields and could be dangerous to public and grazing livestock. To minimise these risks the continuous application of urban waste needs to be monitored in order to ensure that heavy metal and overall nutrient concentrations do not reach toxic levels and do not damage the environment through leaching and eutrophication.

Urban waste also contains organic compounds such as dyes, inks, pesticides, and solvents that are often found in commercial and industrial sludge. These pathogens have been shown to cause genetic damage, while others, such as bacteria, protozoa, and viruses can cause salmonellosis, amoebic dysentery, and infectious hepatitis. Untreated urban waste can put these pathogens in contact with fruits and vegetables. One option is to compost the sludge.

Composting concentrates nutrients and helps to kill disease-causing organisms, slow the release of nitrogen that might otherwise percolate into groundwater, and eliminate aesthetically objectionable odours.

Another option is to use ionising radiation to kill pathogens in and on food without affecting taste. Despite some public concern about the safety of food irradiation, the technique is likely to be adopted more fully in the future in order to protect public health, improve the shelf-life of food, and make it more feasible to apply treated, nutrient-rich urban waste to farmland.

Currently, effective use of urban waste is hampered by its high water content, bulkiness, distance from rural areas, contamination with nondecomposable household items, and high handling, storage, transport, and application costs. However, given the cost and the lack of availability of inorganic fertlisers in some areas, the relative abundance and benefit of urban waste as a soil amendment, and the rising cost of environmentally safe waste disposal, economies may make urban waste an appropriate fertilizer choice in areas where agricultural lands are near urban centres.

Alternative sources of inorganic fertlisers will also be required in the future, particularly in those parts of Africa where the fertility of the soil needs to be rebuilt and high costs and supply constraints limit the application of fertilizer. Soil infertility (particularly phosphorus deficiency) in parts of semiarid West Africa, for example, limits crop production more than the lack of moisture. Phosphorus application of 15-20 kilograms per hectare can substantially improve crop yields.

Medium-reactive and partially acidulated, less-reactive phosphate rock found in Mali, Niger, and Senegal are as agronomically effective as commercial superphosphate fertlisers. Low-cost technology needs to be developed so that phosphate fertilizer can be locally produced from these and other untouched or currently uneconomical phosphate rock reserves.

Where phosphate deficiency is severe in Sub-Saharan Africa, government assistance in developing lowcost technology and in applying phosphate fertilizer should be evaluated. Because phosphorus and phosphorus rock bind to the soil and thereby reduce the opportunity for leaching, and because these fertlisers release nutrients slowly over time, their use to preserve the long-term productive capacity of the land should be considered more of a capital investment than a subsidy or an environmentally undesirable government intervention.

Internal nutrient sources

Although new sources of nutrients can be developed, genetic engineering offers the potential for plants themselves to generate some of the nutrients they require through nitrogen fixation. In this process, rhizobium bacteria infect, invade, and draw energy from leguminous plants, and in return the bacteria convert and store atmospheric nitrogen in a form that the plant can use for growth. Besides helping the plants themselves, cereals grown in rotation with leguminous plants can absorb the nitrates released from the decaying roots and nodules of the leguminous plants. Experiments have shown that rice-legume rotations can result in a 30 percent reduction in chemical fertilizer use.

Genetic research has begun to identify the genes responsible for such nitrogen fixation and assimilation. Further research offers the opportunity of altering or developing microorganisms that can fix nitrogen in nonleguminous plants, such as cereals. As with leguminous plants, plant nitrogen needs could be partially met by the plant itself, such that farmers would then simply need to topup crops with inorganic nitrogen fertlisers.

The task is considerable. Some 17 genes code the enzymes involved in nitrogen fixation. Since these genes, as well as the genes necessary for nodule formation, need to be transferred, the process is complex and its realisation will be costly. Furthermore, because the amount of energy required to fix 150 kilograms of nitrogen per hectare could reduce wheat yields by 20-30 percent, an appropriate balance needs to be found between the nutrient-supply-enhancing benefits of nitrogen fixation and the potential reduction in yields.

REFERENCES

Alston. J. M., P. G. Pardey, and V. H. Smith, "Financing agricultural R&D in rich countries: What's happening and why", *Australian Journal of Agricultural and Resource* 42(1), 1998.

Ange, A. L., *Integrated plant nutrition system* on farm and soil fertility management: The *major concepts*, Rome: FAO, 1992.

Benneh, George, *Toward sustainable agriculture in Sub-Saharan Africa: Issues and strategies*, IFPRI Lecture Series No. 4. Washington, DC: International Food Policy Research Institute, 1997.

Bockman, O. C., O. Kaarstad, O. H. Lie, and I. Richards, *Agriculture and fertilizers: Fertilizers in perspective*, Oslo: Norsk Hydro, 1990.

Conway, G. R., and J. N. Pretty, *Unwelcome harvest: Agriculture and pollution*, London: Earthscan Publications Ltd., 1991.

Coster, R., "Alleviating fertilizer supply constraints in West Africa", In *Alleviating soil fertility constraints to increased crop production in West Africa*, ed. A. Uzo Mokwunye, Dordrecht: Kluwer Academic Publishers, 1991.

Donovan, G., and F. Casey, *Soil fertility management in Sub-Saharan Africa*, World Bank Technical Paper No. 408. Washington, DC: World Bank,. 1998.

Dregne, H. E., "Erosion and soil productivity in Africa", *Journal of Soil and Water Conservation* 45:431-436, 1990.

Freney, J. R., "Efficient use of fertilizer nitrogen by crops", In *Appropriate use of fertilizers in Asia and the Pacific*, ed. S. Ahmed. Taipei: Food and Fertilizer Technology Center, 1996.

Gill, G. J., *Major natural resource management concerns in South Asia*, Food, Agriculture, and the Environment Discussion Paper 8. Washington, DC: IFPRI, 1995.

4

Sustainable Agriculture and Fertiliser Use

Sustainable agriculture integrates three main goals—environmental health, economic profitability, and social and economic equity. A variety of philosophies, policies and practices have contributed to these goals. People in many different capacities, from farmers to consumers, have shared this vision and contributed to it. Despite the diversity of people and perspectives, the following themes commonly weave through definitions of sustainable agriculture. Sustainability rests on the principle that we must meet the needs of the present without compromising the ability of future generations to meet their own needs.

Therefore, stewardship of both natural and human resources is of prime importance. Stewardship of human resources includes consideration of social responsibilities such as working and living conditions of labourers, the needs of rural communities, and consumer health and safety both in the present and the future. Stewardship of land and natural resources involves maintaining or enhancing this vital resource base for the long term. A systems perspective is essential to understanding sustainability.

The system is envisioned in its broadest sense, from the individual farm, to the local ecosystem, and to communities affected by this farming system both locally and globally.

An emphasis on the system allows a larger and more thorough view of the consequences of farming practices on both human communities and the environment. A systems approach gives us the tools to explore the interconnections between farming and other aspects of the environment. A systems approach also implies interdisciplinary efforts in research and education. This requires not only the input of researchers from various disciplines, but also farmers, farmworkers, consumers, policymakers and others.

Making the transition to sustainable agriculture is a process. For farmers, the transition to sustainable agriculture normally requires a series of small, realistic steps. Family economics and personal goals influence how fast or how far participants can go in the transition. It is important to realise that each small decision can make a difference and contribute to advancing the entire system further on the "sustainable agriculture continuum." The key to moving forward is the will to take the next step.

Finally, it is important to point out that reaching toward the goal of sustainable agriculture is the responsibility of all participants in the system, including farmers, labourers, policymakers, researchers, retailers, and consumers. Each group has its own part to play, its own unique contribution to make to strengthen the sustainable agriculture community.

FARMING AND NATURAL RESOURCES

Water

When the production of food and fibber degrades the natural resource base, the ability of future generations to produce and flourish decreases. The decline of ancient civilisations in Mesopotamia, the Mediterranean region, Pre-Columbian southwest U.S. and Central America is believed to have been strongly influenced by natural resource degradation from non-sustainable farming and forestry practices. Water is the principal resource that has helped agriculture and society to prosper, and it has been a major limiting factor when mismanaged.

Water Supply and Use

In California, an extensive water storage and transfer system has been established which has allowed crop production to expand to very arid regions. In drought years, limited surface water supplies have prompted overdraft of groundwater and consequent intrusion of salt water, or permanent collapse of aquifers. Periodic droughts, some lasting up to 50 years, have occurred in California. Several steps should be taken to develop drought-resistant farming systems even in "normal" years, including both policy and management actions:

— improving water conservation and storage measures,
— providing incentives for selection of drought-tolerant crop species,
— using reduced-volume irrigation systems,
— managing crops to reduce water loss, or
— not planting at all.

Water Quality

The most important issues related to water quality involve salinisation and contamination of ground and surface waters by pesticides, nitrates and selenium. Salinity has become a problem wherever water of even relatively low salt content is used on shallow soils in arid regions and/or where the water table is near the root zone of crops. Tile drainage can remove the water and salts, but the disposal of the salts and other contaminants may negatively affect the environment depending upon where they are deposited. Temporary solutions include the use of salt-tolerant crops, low-volume irrigation, and various management techniques to minimise the effects of salts on crops.

In the long-term, some farmland may need to be removed from production or converted to other uses. Other uses include conversion of row crop land to production of drought-tolerant forages, the restoration of wildlife habitat or the use of agroforestry to minimise the impacts of salinity and high water tables.

Wildlife

Another way in which agriculture affects water resources is through the destruction of riparian habitats within watersheds. The conversion of wild habitat to agricultural land reduces fish and wildlife through erosion and sedimentation, the effects of pesticides, removal of riparian plants, and the diversion of water. The plant diversity in and around both riparian and agricultural areas should be maintained in order to support a diversity of wildlife. This diversity will enhance natural ecosystems and could aid in agricultural pest management.

Energy

Modern agriculture is heavily dependent on non-renewable energy sources, especially petroleum. The continued use of these energy sources cannot be sustained indefinitely, yet to abruptly abandon the reliance on them would be economically catastrophic. However, a sudden cutoff in energy supply would be equally disruptive. In sustainable agricultural systems, there is reduced reliance on non-renewable energy sources and a substitution of renewable sources or labour to the extent that is economically feasible.

Air

Many agricultural activities affect air quality. These include smoke from agricultural burning; dust from tillage, traffic and harvest; pesticide drift from spraying; and nitrous oxide emissions from the use of nitrogen fertiliser. Options to improve air quality include incorporating crop residue into the soil, using appropriate levels of tillage, and planting wind breaks, cover crops or strips of native perennial grasses to reduce dust.

Soil

Soil erosion continues to be a serious threat to the continued ability to produce adequate food. Numerous practices have been developed to keep soil in place, which include reducing or eliminating tillage, managing irrigation to reduce runoff, and keeping the soil covered with plants or mulch.

PLANT PRODUCTION PRACTICES

Sustainable production practices involve a variety of approaches. Specific strategies must take into account topography, soil characteristics, climate, pests, local availability of inputs and the individual grower's goals. Despite the site-specific and individual nature of sustainable agriculture, several general principles can be applied to help growers select appropriate management practices:

— Selection of species and varieties that are well suited to the site and to conditions on the farm;

— Diversification of crops (including livestock) and cultural practices to enhance the biological and economic stability of the farm;

— Management of the soil to enhance and protect soil quality;

— Efficient and humane use of inputs; and

— Consideration of farmers' goals and lifestyle choices.

Selection of site, species and variety. Preventive strategies, adopted early, can reduce inputs and help establish a sustainable production system. When possible, pest-resistant crops should be selected which are tolerant of existing soil or site conditions. When site selection is an option, factors such as soil type and depth, previous crop history, and location (e.g. climate, topography) should be taken into account before planting.

Diversity

Diversified farms are usually more economically and ecologically resilient. While monoculture farming has advantages in terms of efficiency and ease of management, the loss of the crop in any one year could put a farm out of business and/or seriously disrupt the stability of a community dependent on that crop. By growing a variety of crops, farmers spread economic risk and are less susceptible to the radical price fluctuations associated with changes in supply and demand.

Properly managed, diversity can also buffer a farm in a biological sense. For example, in annual cropping systems, crop rotation can be used to suppress weeds, pathogens and insect pests. Also, cover crops can have stabilising effects on the agroecosystem by holding soil and nutrients in place, conserving soil moisture with mowed or standing dead mulches, and by increasing the water infiltration rate and soil water holding capacity.

Cover crops in orchards and vineyards can buffer the system against pest infestations by increasing beneficial arthropod population and can therefore reduce the need for chemical inputs. Using a variety of cover crops is also important in order to protect against the failure of a particular species to grow and to attract and sustain a wide range of beneficial arthropods.

Optimum diversity may be obtained by integrating both crops and livestock in the same farming operation. This was the common practice for centuries until the mid-1900s when technology, government policy and economics compelled farms to become more specialised. Mixed crop and livestock operations have several advantages. First, growing row crops only on more level land and pasture or forages on steeper slopes will reduce soil erosion. Second, pasture and forage crops in rotation enhance soil quality and reduce erosion; livestock manure, in turn, contributes to soil fertility. Third, livestock can buffer the negative impacts of low rainfall periods by consuming crop residue that in "plant only" systems would have been considered crop failures. Finally, feeding and marketing are flexible in animal production systems. This can help cushion farmers against trade and price fluctuations and, in conjunction with cropping operations, make more efficient use of farm labour.

Soil Management

A common philosophy among sustainable agriculture practitioners is that a "healthy" soil is a key component of sustainability; that is, a healthy soil will produce healthy crop plants that have optimum vigor and are less susceptible to pests. While many crops have key pests that attack even the healthiest of plants, proper soil, water and nutrient management can help prevent some pest problems brought on by crop stress or nutrient imbalance.

In sustainable systems, the soil is viewed as a fragile and living medium that must be protected and nurtured to ensure its long-term productivity and stability. Methods to protect and enhance the productivity of the soil include using cover crops, compost and/or manures, reducing tillage, avoiding traffic on wet soils, and maintaining soil cover with plants and/or mulches. Conditions in most California soils (warm, irrigated, and tilled) do not favour the buildup of organic matter. Regular additions of organic matter or the use of cover crops can increase soil aggregate stability, soil tilth, and diversity of soil microbial life.

Efficient use of Inputs

Many inputs and practices used by conventional farmers

are also used in sustainable agriculture. Sustainable farmers, however, maximise reliance on natural, renewable, and on-farm inputs. Equally important are the environmental, social, and economic impacts of a particular strategy. Converting to sustainable practices does not mean simple input substitution. Frequently, it substitutes enhanced management and scientific knowledge for conventional inputs, especially chemical inputs that harm the environment on farms and in rural communities. The goal is to develop efficient, biological systems which do not need high levels of material inputs.

Growers frequently ask if synthetic chemicals are appropriate in a sustainable farming system. Sustainable approaches are those that are the least toxic and least energy intensive, and yet maintain productivity and profitability. Preventive strategies and other alternatives should be employed before using chemical inputs from any source. However, there may be situations where the use of synthetic chemicals would be more "sustainable" than a strictly nonchemical approach or an approach using toxic "organic" chemicals. For example, one grape grower switched from tillage to a few applications of a broad spectrum contact herbicide in the vine row. This approach may use less energy and may compact the soil less than numerous passes with a cultivator or mower.

Consideration of farmer goals and lifestyle choices. Management decisions should reflect not only environmental and broad social considerations, but also individual goals and lifestyle choices. For example, adoption of some technologies or practices that promise profitability may also require such intensive management that one's lifestyle actually deteriorates. Management decisions that promote sustainability, nourish the environment, the community and the individual.

ANIMAL PRODUCTION PRACTICES

In the early part of this century, most farms integrated both crop and livestock operations. Indeed, the two were highly complementary both biologically and economically. The current picture has changed quite drastically since then. Crop and animal producers now are still dependent on one another to some degree, but the integration now most commonly takes place at a higher level—between farmers, through intermediaries, rather than within the farm itself. This is the result of a trend toward separation and specialisation of crop and animal production systems. Despite this trend, there are still many farmers, particularly in the Midwest and Northeastern U.S. that integrate crop and animal systems—either on dairy farms, or with range cattle, sheep or hog operations. Even with the growing specialisation of livestock and crop producers, many of the principles outlined in the crop production section apply to both groups. The actual management practices will, of course, be quite different.

Management Planning

Including livestock in the farming system increases the complexity of biological and economic relationships. The mobility of the stock, daily feeding, health concerns, breeding operations, seasonal feed and forage sources, and complex marketing are sources of this complexity. Therefore, a successful ranch plan should include enterprise calendars of operations, stock flows, forage flows, labour needs, herd production records and land use plans to give the manager control and a means of monitoring progress toward goals.

Animal Selection

The animal enterprise must be appropriate for the farm or ranch resources. Farm capabilities and constraints such as feed and forage sources, landscape, climate and skill of the manager must be considered in selecting which animals to produce. For example, ruminant animals can be raised on a variety of feed sources including range and pasture, cultivated forage, cover crops, shrubs, weeds, and crop residues. There is a wide range of breeds available in each of the major ruminant species, i.e., cattle, sheep and goats. Hardier breeds that, in general, have lower growth and milk production potential, are better adapted to less favourable environments with sparse or highly seasonal forage growth.

Animal Nutrition

Feed costs are the largest single variable cost in any livestock operation. While most of the feed may come from other enterprises on the ranch, some purchased feed is usually imported from off the farm. Feed costs can be kept to a minimum by monitoring animal condition and performance and understanding seasonal variations in feed and forage quality on the farm. Determining the optimal use of farm-generated by-products is an important challenge of diversified farming.

Reproduction

Use of quality germplasm to improve herd performance is another key to sustainability. In combination with good genetic stock, adapting the reproduction season to fit the climate and sources of feed and forage reduce health problems and feed costs.

Herd Health

Animal health greatly influences reproductive success and weight gains, two key aspects of successful livestock production. Unhealthy stock waste feed and require additional labour. A herd health programme is critical to sustainable livestock production.

Grazing Management

Most adverse environmental impacts associated with grazing can be prevented or mitigated with proper grazing management. First, the number of stock per unit area (stocking rate) must be correct for the landscape and the forage sources. There will need to be compromises between the convenience of tilling large, unfenced fields and the fencing needs of livestock operations. Use of modern, temporary fencing may provide one practical solution to this dilemma. Second, the long term carrying capacity and the stocking rate must take into account short and long-term droughts. Especially in Mediterranean climates such as in California, properly managed grazing significantly reduces fire hazards by reducing fuel build-up in grasslands and brushlands.

Finally, the manager must achieve sufficient control to reduce overuse in some areas while other areas go unused. Prolonged concentration of stock that results in permanent loss of vegetative cover on uplands or in riparian zones should be avoided. However, small scale loss of vegetative cover around water or feed troughs may be tolerated if surrounding vegetative cover is adequate.

Confined Livestock Production

Animal health and waste management are key issues in confined livestock operations. The moral and ethical debate taking place today regarding animal welfare is particularly intense for confined livestock production systems. Confinement livestock production is increasingly a source of surface and ground water pollutants, particularly where there are large numbers of animals per unit area. Expensive waste management facilities are now a necessary cost of confined production systems. Waste is a problem of almost all operations and must be managed with respect to both the environment and the quality of life in nearby communities. Livestock production systems that disperse stock in pastures so the wastes are not concentrated and do not overwhelm natural nutrient cycling processes have become a subject of renewed interest.

ECONOMIC, SOCIAL AND POLITICAL CONTEXT

In addition to strategies for preserving natural resources and changing production practices, sustainable agriculture requires a commitment to changing public policies, economic institutions, and social values. Strategies for change must take into account the complex, reciprocal and ever-changing relationship between agricultural production and the broader society. The "food system" extends far beyond the farm and involves the interaction of individuals and institutions with contrasting and often competing goals including farmers, researchers, input suppliers, farmworkers, unions, farm advisors, processors, retailers, consumers, and policymakers. Relationships among these actors shift over time as new technologies spawn economic, social and political changes.

A wide diversity of strategies and approaches are necessary to create a more sustainable food system. These will range from specific and concentrated efforts to alter specific policies or practices, to the longer-term tasks of reforming key institutions, rethinking economic priorities, and challenging widely-held social values. Areas of concern where change is most needed include the following:

Food and Agricultural Policy

Existing federal, state and local government policies often impede the goals of sustainable agriculture. New policies are needed to simultaneously promote environmental health, economic profitability, and social and economic equity. For example, commodity and price support programmes could be restructured to allow farmers to realise the full benefits of the productivity gains made possible through alternative practices. Tax and credit policies could be modified to encourage a diverse and decentralised system of family farms rather than corporate concentration and absentee ownership. Government and land grant university research policies could be modified to emphasize the development of sustainable alternatives. Marketing orders and cosmetic standards could be amended to encourage reduced pesticide use. Coalitions must be created to address these policy concerns at the local, regional, and national level.

Land Use

Conversion of agricultural land to urban uses is a particular concern in California, as rapid growth and escalating land values threaten farming on prime soils. Existing farmland conversion patterns often discourage farmers from adopting sustainable practices and a long-term perspective on the value of land. At the same time, the close proximity of newly developed residential areas to farms is increasing the public demand for environmentally safe farming practices. Comprehensive new policies to protect prime soils and regulate development are needed, particularly in California's Central Valley. By helping farmers to adopt practices that reduce chemical use and conserve scarce resources, sustainable agriculture research and education can play a key role in building public support for agricultural land preservation. Educating land use planners and decision-makers about sustainable agriculture is an important priority.

Labour

In California, the conditions of agricultural labour are generally far below accepted social standards and legal protections in other forms of employment. Policies and programmes are needed to address this problem, working toward socially just and safe employment that provides adequate wages, working conditions, health benefits, and chances for economic stability. The needs of migrant labour for year-around employment and adequate housing are a particularly crucial problem needing immediate attention. To be more sustainable over the long-term, labour must be acknowledged and supported by government policies, recognised as important constituents of land grant universities, and carefully considered when assessing the impacts of new technologies and practices.

Rural Community Development

Rural communities in California are currently characterised by economic and environmental deterioration. Many are among the poorest locations in the nation. The reasons for the decline are complex, but changes in farm structure have played a significant role. Sustainable agriculture presents an opportunity to rethink the importance of family farms and rural communities. Economic development policies are needed that encourage more diversified agricultural production on family farms as a foundation for healthy economies in rural communities. In combination with other strategies, sustainable agriculture practices and policies can help foster community institutions that meet employment, educational, health, cultural and spiritual needs.

Consumers and the Food System

Consumers can play a critical role in creating a sustainable food system. Through their purchases, they send strong messages to producers, retailers and others in the system about what they think is important. Food cost and nutritional quality have always influenced consumer choices. The challenge now is to find strategies that broaden consumer perspectives, so that environmental quality, resource use, and social equity issues are also considered in shopping decisions.

At the same time, new policies and institutions must be created to enable producers using sustainable practices to market their goods to a wider public. Coalitions organised around improving the food system are one specific method of creating a dialogue among consumers, retailers, producers and others. These coalitions or other public forums can be important vehicles for clarifying issues, suggesting new policies, increasing mutual trust, and encouraging a long-term view of food production, distribution and consumption.

Use of fertilisers

Sustainable agriculture must produce enough food and fibre to satisfy changing human needs while conserving natural resources, maintaining the quality of environment and ultimately leading to community and gender equity. In the developed countries in Europe and to some extent in USA, chemical fertiliser has been held as a major culprit for environmental pollution, especially the nitrate enrichment of groundwater. However, evidence available also indicates that, animal slurry and septic cess pools largely contribute to nitrate enrichment of groundwater and also to environmental pollution with ammonia and nitrogen oxides (NO_x).

There are other sources of environmental pollution such as exhaust fumes from motor vehicles, flyash from thermal power plants and other industrial effluents. Developing countries such as India, reeling under population pressure with no additional cultivable land are forced to increase their fertiliser consumption, which as of today is much less than the actual crop needs. Nevertheless, we must learn lessons from the ill-effects of overuse of chemical fertiliser by developed countries and use it judiciously with a well-planned integrated plant nutrient supply system.

Sustainability of the agricultural systems is of global concern today and many definitions of sustainable agriculture have become available. The five main components of these definitions are:

— production of enough food and fibre to meet the increasing and changing needs of the people;
— conservation of natural resources;
— maintaining the quality of environment;
— achieving community and gender equity; and
— avoidance of regional imbalances

In agriculture we convert solar energy into chemical energy, i.e. food, feed or fibre. It is common knowledge that this is achieved with the help of chlorophyll in plants. Keeping at optimum other conditions such as solar radiation, temperature, etc. the more the chlorophyll the higher the biomass and it is likely that more food is produced. However, for doing this plants need adequate supply of 13 elements (N, P, K, Ca, Mg, S, Fe, Mn, Zn, Cu, B, Mo, Cl) from soil, which are known as essential plant nutrients. Fertilisers are the chemicals that supply these essential plant nutrients, mostly N, P and K, which are removed by crop plants in the largest quantities.

In one study at New Delhi 28 kg N, 4.4 kg P, 41 kg K, 4.9 kg S, 5.9 kg Ca, 4.2 kg Mg, 400 g Fe, 100 g each of Mn and Zn and 30 g Cu were removed from the soil for each tonne of wheat grain produced. Similarly from the International Rice Research Institute, Philippines it is reported that 10–31 kg N, 1–5 kg P and 8–35 kg K were removed from the soil per tonne of rice grain produced. When crop yields of 5–8 t grain ha^{-1} are taken as with the high yielding varieties of crops, it is not possible for most soils to supply the needed amounts of plant nutrients and that is why fertilisers are needed.

Such heavy removal of plant nutrients from soil leads to depletion of soil fertility, which shows up in crop yield decline and lowered factor productivity. For example, in a long-term study at four research centres rice yields declined by 32% in plots receiving nitrogen and phosphate fertilisation; the decline in unfertilised plots was 57%. From the dryland rainfed areas where yields are low and very little fertiliser applied, a negative balance between crop removal and plant nutrient application of 6.37 million tonnes yr^{-1} ($N + P_2O_5 + K_2O$) in India has been reported. Thus contrary to lowering the rates of fertiliser application, Indian farmers may have to think of fertilising for crop production as well as for soil fertility resilience. Thus fertiliser has been and will continue to be the king input in India's achieving self sufficiency in foodgrain production.

Fertilisers and Sustainability of Agriculture

For the last three decades all over the developed world fingers have been raised on fertiliser, particularly nitrogen, as the number one enemy of sustainable agriculture. One should not be surprised to see this happen if the rates of application are too high such as in Netherlands.

Nitrogen is a mobile nutrient both in plants and soils and considering the fact that its efficiency (nitrogen taken up by the above ground portion of crop expressed as percentage of that applied) varies from 30–40% in rice to 60–80% in other cereals, a sizeable amount could be added to the environment as ammonia by volatilisation from soil surface, nitrous oxide or elemental nitrogen by denitrification which is not restricted to tropical rice regions, but also applies to temperate regions and finally as nitrates by leaching in underground water.

The ammonia going in the atmosphere contributes to acid rains, while N_2O is involved in depletion of ozone layer. What one generally overlooks is that in the case of nitrogen fertilizers we recycle atmospheric nitrogen which is the raw material for ammonia and urea manufacture. Atleast 30–50% of it is converted into human edible food and about one-third is immobilised in the soil and only the rest goes back to atmosphere either as ammonia or as N_2O or N_2 after denitrification of nitrates. Phosphates, which are not so mobile in soil and get fixed in the soil as insoluble compounds could also leach from very light soils and may also move with eroded surface soil to surface waters such as lakes and ponds.

Within the European Union, the average load of nitrogen in 1990 ranged from 35 kg N ha^{-1} in Spain to 195 kg N ha^{-1} in the Netherlands and average loads of phosphate ranged from 17 kg P_2O_5 ha^{-1} in Spain to 57 kg P_2O_5 ha^{-1} in the Belgium. So great has been the concern of European researchers regarding pollution caused by fertilisers (and manures) that it has been referred to as a 'chemical time bomb' and to overcome the explosion of this the need for immediate change in the land management amounting to abandonment of part of the agricultural land has been suggested.

Pollution of Waters

More than any other aspect of environmental degradation, pollution of surface and underground waters due to fertilisers has received greater attention. Water can be broadly classified into four categories:

— non-flowing such as lakes and ponds;

— flowing such as rivers and canals;

— river estuaries and

— underground water.

Both nitrogen and phosphorus eutrophication of lakes and ponds leads to excessive growth of aquatic plants (macrophhytes) and algae (phytoplankton), which deplete the water of oxygen and this may lead to death of fishes and other aquatic animal life. Furthermore, algal and decaying algal and aquatic plant tissue can lead to discolouration and bad odour and this affects recreational and aesthetic water uses. As regards flowing water, phosphorus is the limiting nutrient. When all phosphorus is used, plant growth ceases irrespective of the amount of nitrogen available. Thus only if adequate phosphorus is available, increasing concentration of nitrates will lead to algal and macrophyte growth.

In estuaries, on the other hand, nitrogen controls the growth of algae and macrophytes. In USA nitrates and phosphates are suspected of causing *hypoxia* of the Dead Zone of Gulf of Mexico. Nutrient-enriched run-off from agriculture is also incriminated in the *pfiesteria* problem that killed a large number of fish in the Chakaspeake Bay, USA. The recommended level of nitrogen is 0.1 to 1 mg l^{-1}, while the recommended level of phosphorus is nearly one-tenth of that of nitrogen.

Underground water is the major source of drinking water in India and even in developed countries like USA, 90% of the rural population depends on it for drinking water. The safe limit or MCL (maximum contamination level) established by the US Environmental Protection Agency is 45 mg NO_3–N or 10 mg N l^{-1}. The European Union has fixed MCL limit at 50 mg NO_3–N or 11 mg Nl^{-1}. Levels above this may lead to methaemoglobinaemia or blue baby syndrome. This is due to conversion of haemoglobin to methaemoglobin due to nitrites formed from nitrates ingested with drinking water. Haemoglobin is involved in transport of oxygen in the body, while methaemoglobin cannot and thus the patient suffers from anexia (lack of oxygen). There are some indications that excess nitrates in the human body may react with amines and form nitrosamines that may lead to gastric cancer, but substantive data are missing.

In India the highest rate of fertiliser N is in Punjab, where an increase in shallow well waters from 0.04–6.15 in 1975 to 0.31–13.3 mg NO_3–Nl^{-1} in 1988 has been reported. A recent survey in Delhi reported a range of 26–150 mg

NO_3–Nl^{-1} in shallow well water, which is very high constituting a warning.

Sources of Environmental Pollution

There are a number of sources other than fertilisers which are responsible for environmental degradation. These include livestock and human excrement in rural areas and leaking septic systems, sewage, combustion of fossil fuels in motor cars and other vehicles in urban areas. Industries that use nitrates in their manufacturing processes such as meat curing, production of explosives including fire crackers, heat transfer fluid, etc. may also release substantial nitrates in their effluent water. Precise estimates on all these are not available.

Livestock excrement

Ammonia emissions from dairy farms are one of the major sources of environmental pollution in Europe. Losses of ammonia occur during slurry application, housing, slurry storage and grazing. On an average 30% of nitrogen excreted by farm animals is released to the atmosphere as ammonia and the bulk of it is contributed by urine. Estimates indicate that 50–90% of N contained in urine is urea and 4–41% of N from applied cattle urine may volatilise. The average ammonia emission per unit land area in Western Europe is 12 kg N ha^{-1} yr^{-1} and the highest is 45 kg ha^{-1} yr^{-1}; as a contrast ammonia emission from growing arable crops is estimated at 1–2 kg ha^{-1} yr^{-1}. What is to be noted is that ammonia emitted is not deposited in the same area but transported to other areas depending upon wind direction and velocity.

In India where cowdung is mostly used as cooking fuel, loss of nitrogen could be partly as ammonia during storage of wet dung but mostly as nitrogen oxides (NO_x) when the cowdung cakes are burnt. Nevertheless it does contribute to environmental pollution. We need to measure and generate data on this. Ammonia emission is not the only way the animal excreta contributes to environment. After its application to soil, most of ammonia produced is retained by the soil exchange complex and is gradually nitrified.

Nitrates are then leached to deeper layers and contribute to underground water pollution. From a study in a rural community in California (USA), based on ^{15}N ratio studies it was concluded that much of the nitrate in groundwater came from sewage or manure. As already discussed, under anoxic conditions nitrates are denitrified to nitrous oxides or elemental N (generally referred to as NO_x). In UK limits of organic manures have been fixed at 250 kg N ha^{-1} on grassland and 210 kg N ha^{-1} on arable land in nitrate vulnerable zones (NVZ); the figure for arable land will be reduced to 170 kg N ha^{-1} from 19 December 2002.

Septic systems

Septic systems and cesspools (a pit or a hole in which sewage collects) in heavily populated areas such as Long Island in USA have been major sources of nitrate in groundwater for decades and public supply wells in the upper glacial aquifer of Nassau county were abandoned in 1949 because of nitrate contamination. Katz *et al.* estimated that 11.6 metric t N km^{-2} were contributed by cesspools and septic systems in the Nassau county in 1975.

Thermal power houses

While phosphate fertilisers have been blamed for contaminating soil with heavy metals, thermal power plants could also do it in their vicinity by generation and disposal of flyash. Of the total power production in India, 65.3% (during 1989–90) came from thermal power production. About 35 million tonnes (MT) of flyash was produced from this and its value may rise to 125 MT by the turn of the century. Considerable amount of flyash can be emitted into surrounding environments and the main method of disposal is by mixing the ash with water, the resultant slurry is then pumped to disposal ponds. Heavy metal composition of fly-ash from the Indraprastha and Rajghat Power Houses at New Delhi and the range of these in the soil in their vicinity. In addition to heavy metals, coal also contains some nitrogen and most of it will be lost to atmosphere as NO_x, which is yet another addition to environmental pollution.

Fossil fuels

Burning of fossil fuels in vehicles also contributes to NO_x to the atmosphere. In a recent study in US, 24–29 tonnes of NH_3 day^{-1} were emitted from a vehicle fleet in the South Coast Air Basin that surrounds Los Angeles. These emissions were from vehicles fitted with three-way catalytic converters designed to reduce NO_x to N_2 and O_2. More studies are needed to find out the accurate global emissions of NO_x and ammonia from vehicles.

Biological nitrogen fixation

Biological nitrogen fixation (BNF) is considered as one of the major ways by which the crop needs of fertiliser nitrogen can be partly (30–60 kg N ha^{-1}) met; the rates for green manuring could be 80–185 kg N ha^{-1}. There can be no two opinions about it, and infact legumes as restorer of soil fertility have been known since ages.

Estimates of total annual terrestrial BNF range from 139 to 170 MT with symbiotic N fixation in arable farming accounting for 35–44 MT N, and another 45 MT from permanent pastures and the rest from non-symbiotic N fixation. In comparison, only 91 MT of fertiliser N was globally manufactured in 1996–97; which is only 53–65% of biologically fixed N. The point to consider is that nitrogen fixed by BNF also enters the soil nitrogen pool and while some of it is available to associated crop, most of it is made available to the succeeding crop.

Biologically-fixed N is much more readily available to plants than the native soil organic N, nevertheless, a part of it is also subject to loss. For example, a loss of 28–29% of nitrogen fixed by *Crotolaria* sp and *Sesbania* sp has been reported when these legumes were green-manured to rice, as compared to 36% from urea nitrogen.

Similarly in maize loss of *Leucaena* nitrogen was 25–30%, while that from ammonium sulphate was 14–35%. Thus part of biologically-fixed nitrogen may also contribute to environmental pollution including nitrate enrichment of groundwater. BNF is reported to contribute as much as 13.5% of total N_2O emission to the atmosphere, which was equal to the amount contributed by fertilisers.

India's Foodgrain Needs

In 1951, we were only 361 million people and by 2000 AD, another year from now, we will be 1004.5 million, nearly trebling the population in the second half of twentieth century. A well-planned and concerted effort by agricultural scientists, extension workers, farmers and the government price support policies have made 'green revolution' realisable by increasing grain production at a rate faster than the increase in population. This has kept us away from hunger despite some very bad drought years. Hungry people can hardly think of a quality environment.

Three major inputs that have made it possible are: seeds of high yielding varieties of crops, fertiliser and increased irrigation facilities. But then it is not all over and although there are good signs of decline in the rate of population increase from 2.14 to 1.70%, we are still likely to add another 420.5 million people in the next 20 years by 2020 AD, that is about 21 million people each year. This will certainly increase the food demand.

The total foodgrain demand by 2020 is estimated at 294 MT. Thus by 2020 we need to produce about 100 MT of additional food grain yr^{-1} from the same or even less area. We have no choice but to increase the fertiliser application. During 1980–90 there has been 3–4% decrease in fertiliser nitrogen consumption in Europe and USA, while in Asia it has increased by 74.4%. In India the current fertiliser consumption is much below the mark. Only 19 out of 437 districts in India consumed more than 200 kg $N + P_2O_5 + K_2O$ ha^{-1} in 1996–97, while 176 districts consumed 50 kg ha^{-1} or less.

Social scientists are quite right in pointing out regional imbalances and lack of equity among communities due to modernisation of agriculture, but fertilisers do not come in the way of achieving either of these social goals. There are other factors responsible such as lack of infrastructure, lack of small scale agro-industry and the need for appropriate government policies, etc.

Only a carefully planned sustainable agriculture can provide a sustainable livelihood security for the poor, which should be the foundation for all development programmes. Gandhiji called such an approach to the development as 'Antyodaya' model, which even foreigners are talking about and in its operational sense implies that the priorities in development should be measured by their potential benefit to the poorest sections of the community. In this exercise food of course will come as the first priority and in countries like India it can be produced only by the judicious use of fertilisers.

As regards environmental problems we have to find out ways to overcome these and reduce pollution hazards. A number of techniques are used by agricultural scientists for reducing nitrate leaching losses. These include growing high-yielding crop varieties, deep placement and split application of nitrogen fertilisers, use of ammonium or amide fertilisers and adopting an integrated plant nutrient system (IPNS).

Dudal and Roy defined IPNS as an approach which adapts plant nutrition to a specific farming system and particular yield targets, the physical resource base, the available plant nutrient sources and the socio-economic background. The sources of plant nutrients may be mineral fertilisers and/or biological nitrogen fixation and/or organic materials depending upon a particular location.

The FAO–IFFCO International Seminar on IPNS recommended that IPNS should be science-based, associating agronomy, ecology and social sciences. It should use a farming system approach and not limit itself to cropping systems only. It should address both increased productivity and profitability and integrate maintenance and rehabilitation of natural resources.

REFERENCES

Alston. J. M., P. G. Pardey, and V. H. Smith, "Financing agricultural R&D in rich countries: What's happening and why", *Australian Journal of Agricultural and Resource* 42(1), 1998.

Ange, A. L., *Integrated plant nutrition system* on farm and soil fertility management: The *major concepts*, Rome: FAO, 1992.

Benneh, George, *Toward sustainable agriculture in Sub-Saharan Africa: Issues and strategies*, IFPRI Lecture Series No. 4. Washington, DC: International Food Policy Research Institute, 1997.

Bockman, O. C., O. Kaarstad, O. H. Lie, and I. Richards, *Agriculture and fertilizers: Fertilizers in perspective*, Oslo: Norsk Hydro, 1990.

Conway, G. R., and J. N. Pretty, *Unwelcome harvest: Agriculture and pollution*, London: Earthscan Publications Ltd., 1991.

Coster, R., "Alleviating fertilizer supply constraints in West Africa", In *Alleviating soil fertility constraints to increased crop production in West Africa*, ed. A. Uzo Mokwunye, Dordrecht: Kluwer Academic Publishers, 1991.

Donovan, G., and F. Casey, *Soil fertility management in Sub-Saharan Africa*, World Bank Technical Paper No. 408. Washington, DC: World Bank,. 1998.

Dregne, H. E., "Erosion and soil productivity in Africa", *Journal of Soil and Water Conservation* 45:431-436, 1990.

Freney, J. R., "Efficient use of fertilizer nitrogen by crops", In *Appropriate use of fertilizers in Asia and the Pacific*, ed. S. Ahmed. Taipei: Food and Fertilizer Technology Center, 1996.

Gill, G. J., *Major natural resource management concerns in South Asia*, Food, Agriculture, and the Environment Discussion Paper 8. Washington, DC: IFPRI, 1995.

5

Role of Soil Organic Matter in Sustainable Agriculture

The quantity of organic matter in a soil is a major indicator of its quality. The amount of young, biologically active organic matter is important because it helps feed crops, maintains soil structure and improves the overall quality of the soil. Under the current farm practices in the country this active soil organic matter is depleted by narrow rotations that focus on cash crops, leaving little room for soilbuilding crops. Sustainable farming, however, forms the important basis for long-term profitability because it maintains quality soils that can provide long-term stable yields.

To achieve soil quality we should treat the organic matter like a bank account. A bank account lets us deposit, save and withdraw something we value. For sustainability it is important to deposit in the account of active organic matter in the soil on a regular basis. Thereby, we build cultural fertility. Saving something for the future will most likely pay off through less disease pressure and stable yields, which will contribute to a more stable income for the farmer over time.

Soil organic matter is only a small part of the total soil. Organic matter comprises plant and animal residues in different stages of decomposition (including substances no longer identifiable). The more stabile, complex portion of the organic matter found in the soil is generally referred to as soil humus.

Although soil organic matter comprises only a small fraction of the total mass of most soils, this dynamic soil component influences soil properties to an extent far out of proportion to the small quantity present. Maintaining a sufficiently high level of quality organic matter is therefore one of the most critical objectives of soil management and really important for sustainable farmers as they depend to a high extent on their soil organic matter levels to provide sufficient nutrients to their crops.

Therefore careful soil management is essential for measuring up to the high expectations and production levels of modern farming. This includes improving physical (structure), chemical (nutrient availability) and biological (soil flora and fauna) properties of soils. Crops show the effects of good organic matter management. They are more likely to resist weeds and disease problems in a healthy soil. Mistakes with maintaining soil structure or inadequately fertilisation will be visible in a standing crop.

SOIL ORGANIC MATTER

In general soil organic matter has a balancing effect on soils. By improving soil structure, organic matter causes a better distribution of water, air and solid particles in soils. It also improves the water infiltration, the water holding capacity and drainage, aeration, and root penetration. This means more water will be available to a crop in periods of drought, and runoff or erosion will be prevented. Organic matter forms stable bonds with mineral soil particles to hold the crumbs in a soil together. Crumbs or aggregates are also held together by lime and glue-like substances produced by soil organisms. The natural binding of humus molecules and clay particles results in a further improvement of soil structure.

On loamy and clay soils organic matter reduces the cohesion and stickiness of clods, making these soils easier to cultivate. On sandy soils organic matter improves the cohesion of particles improving the aggregate stability of these soil. The ideal crumb structure of a soil is built by the life processes of the soil organisms. These soil organisms mix organic matter and soil particles and excrete the glue-like substances that bind them together.

The soil organisms get their essential food from organic matter. This is the key for a good soil structure. Soil structure coupled with adequate supplies of active organic matter may stimulate root growth. Under these conditions, nutrients and oxygen might be more easily available to the roots. A less stressed rooting system is also more likely to resist diseases. A well-rooted soil in turn is likely to hold soil particles together.

Crop Yields

In a sustainable agriculture organic matter is the key to making essential nutrients available to crops. Poor soil structure and an idle soil life can be disguised for a time in conventional agriculture by excessive fertiliser and pesticide use. Within sustainable practices a long-term strategy with a sound crop rotation and balanced fertilisation is needed to build a healthy soil over the years. During decomposition or mineralisation organic matter is modified by soil organisms into simpler compounds. In this process, part of the carbon that makes up organic matter is released as carbon dioxide. Nutrients that were tied up in the organic matter are also released. Nutrients like nitrogen, phosphorus, sulfur and micronutrients come available during this process. Potassium will be mainly available in an inorganic form.

Microorganisms can create new, complex, humus-like structures that are more resistant to further decomposition. Microorganisms can use nitrogen, phosphorus and potassium to form new compounds, or plant roots can take up these nutrients. But nutrients in the soil moisture might be lost during periods of abundant rainfall through leaching. Increasing the quantity of active organic matter in the soil will increase the amount of microbes living in the soil and these microbes will help to absorb excess nitrate that is present in the soil. Nutrient losses from the sustainable system through leaching are also limited by organic matter since organic matter accounts for between 20 and 90 percent of the capacity of the soil to absorb nutrients.

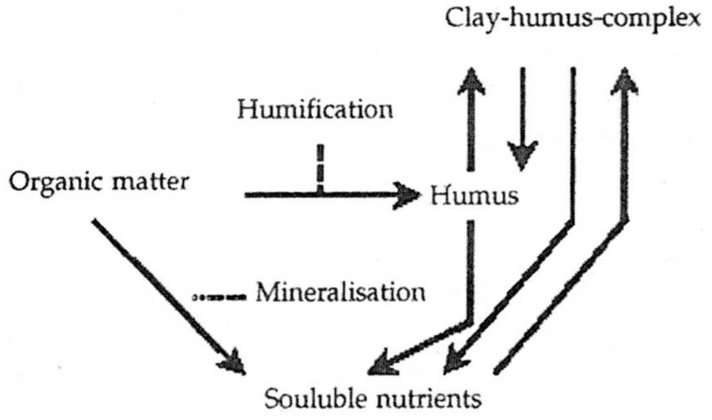

Figure 1. Processes in organic matter formation and turnover

Regular soil analysis provides information on soil pH, phosphorus, potassium, magnesium and probably micronutrient levels. But, often it is difficult for farmers to interpret soil tests and integrate their results into a rational programme for organic management. Advice that is based on experience with conventional farming systems may recommend rates of application of nitrogen containing fertilisers or lime that are too high.

Soil tests need to be reinterpreted based on ecological insights and experience. In a sustainable farming system the fertilisation of the soil is actually a more important principle than the fertilisation of the crop. The most important source of nitrogen comes from mineralisation of crop and manure residues and the active organic matter fraction in the soil. Nitrogen should not be a limiting factor in a sustainable system if there are sufficient quantities of nitrogen fixing legumes in the rotation. Free-living microorganisms can also fix atmospheric nitrogen. Nitrogen washes out of the atmosphere with rain and snow and also makes a contribution. The timely mineralisation of nitrogen from organic matter is dependent on adequate moisture and warmth.

Soil life and Diseases

Organic matter is the source of food for a whole web of organisms in the soils. These organisms in turn are able to give the soil its structure, mobilise nutrients, bind atmospheric nitrogen and suppress soil borne pathogens (Table 1). Organic matter feeds earthworms that in turn improve the aggregate stability and aeration of a soil.

Table 1 The underground animal stock

Component	*typical numbers or length (in one handful of soil)*	*typical biomass (pounds/acre)*
plant roots	60 - 150 in. (annual crops) 1,500-3,000 in (perennial grasses)	3,000 (annual crop) 15,000 (perennial grasses)
bacteria	300 million - 50 billion	400 - 4,000
fungi	0.5 - 100 million	500 - 5,000
actinomycetes	100 million - 2 billion	400 - 4,000
nematodes	1,000 - 10,000	5 - 50
protozoa	100,000 - 50 million	5 - 100
arthropods	100 - 1,000	1 - 10
earthworms	0 - 2	10 - 40

The active soil life is breathing and releasing carbon dioxide (CO_2). A principal prerequisite for microbial soil life is the availability of organic matter. Organic matter stimulates a balanced community of soil organisms. This community can suppress soil borne diseases.

HEALTHY SOIL

Residues and manures are slowly broken into smaller and smaller pieces. As this occurs they change their chemical characteristics and rate of turnover, becoming essentially mixed digestion products from soil organisms and plants. The different stages in the decomposition process can be considered from the standpoint of qualitatively different fractions of organic matter. Depending on their age, size and chemical composition they will have quite different characteristics, which in turn will determine the characteristics of a soil. Perhaps the most useful approach to *defining soil organic matter quality* is to recognise these different fractions of organic matter and how they vary in their susceptibility to microbial metabolism.

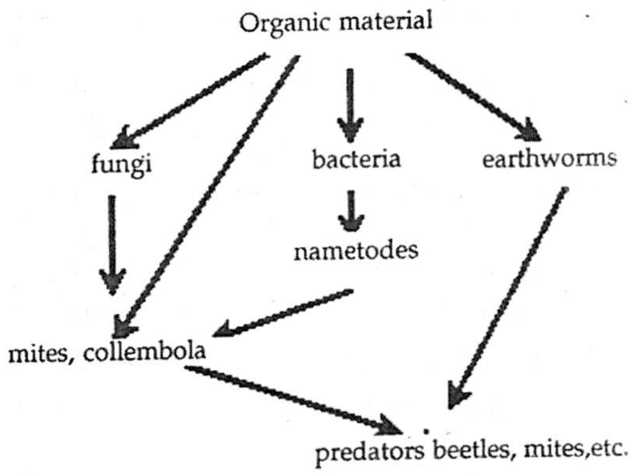

Figure 2. A food web based on organic matter

Components include living biomass, fine particulate

detritus, most of the polysaccharides and nonhumic and other labile fractions. It is characterised by a high C/N ratio. This fraction is comprised most of the readily available food and mineralisable nitrogen. Adding fresh plant material and animal residues increases the quantity of labile organic matter, but it is also rapidly lost when additions are reduced or tillage is intensified. This fraction comprises less than 10 percent of the soil organic matter.

The *passive* fraction of soil organic matter consists of very stable materials remaining in the soil for hundreds or even thousands of years. This fraction includes most of the humus physically protected in clay-humus complexes. This fraction accounts for 50 to 90 percent of the organic mater in most soils, and its quantity is increased or diminished only slowly. This fraction is responsible for much of the cation exchange capacity and water-holding capacity contributed to soil organic matter.

The active fraction is an intermediate state of soil organic matter between the labile and passive fractions. It makes up 20 to 50 percent of the total soil organic matter. This fraction includes very finely divided plant tissue, high in lignin, and other slowly decomposable and chemically resistant components. The turnover of this fraction occurs typically in about two decades.

The active organic matter is very important for maintaining the soil's fertility. It is used by microorganisms to bind aggregates in the soil together to give good structure. Manures and types of perennial forage are important for maintaining the quantities of active organic matter in the soil. The active fraction is an important source of mineralisable nitrogen and other plant nutrients. But it has also beneficial effects on the structure and stability that gives soil, good tilth leading to enhanced water infiltration, resistance to erosion, and ease of tillage.

Studies have shown that different fractions of soil organic matter play quite different roles in soil management and in the carbon cycle. In many cases, changes in soil management result in relatively small effects on soil organic matter, but have pronounced effects on certain soil properties and aspects of soil productivity attributed to soil organic matter.

Table 2. Diseased of vegetable crops that have been at least partially controlled by the use of organic manures

Disease	*Crop*
Pythium root rot	beets, cabbage, lettuce, peas, radishes, spinach, tomatoes, turnips
Rhizoctonia root rot	asparagus, beans, cabbage, carrots, peas, rhubarb, strawberries
Fusarium root rot and Fusarium wilt	asparagus, beans, cabbage, cucumbers, lettuce, peas, peppers, tomatoes
Verticillium	cucumbers, horseradish, peppers
Scab (Streptomyces)	potatoes

A method is available to isolate the active fraction in the soils. Cambardella and Elliott isolated a particulate organic matter fraction (POM) that they suggest corresponds to the characteristics of the active (intermediate) pool in soil organic matter models. Scanning electron microscopy suggests that the POM fraction is composed mostly of root fragments in various stages of decomposition.

Experiments showed that 20 years of wheat cropping reduced the POM fraction from about 40 percent in native prairie to 18 to 25 percent in cultivated soils. A 30-year crop rotation experiment in Moldova shows that it is largely the manure that contributes to POM formation. In addition, a sound rotation that includes legumes is necessary.

Quality of the Residues and their

Formation of organic matter in the soils is largely

determined by the amount and quality of the residues that are incorporated into the soil. Within conventional agriculture much research and focus have been oriented toward increasing yields. For some crops this has meant the amount of residues was increased as well (Table 3).

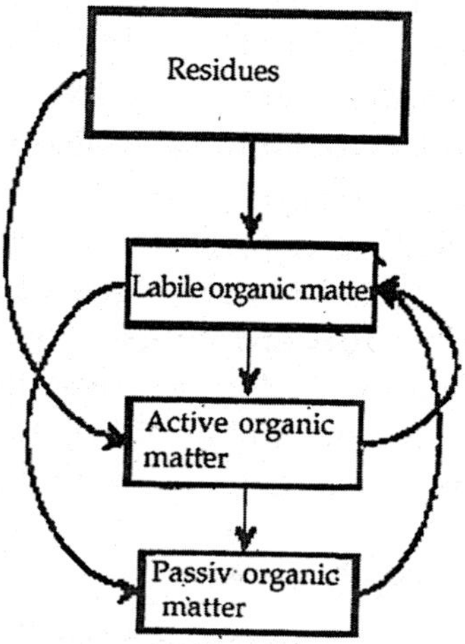

Figure 3: Organic matter quality may result from the distribution of different fraction of organic matter.

The question remaining is whether these residues and the management accompanying them are enough to keep a balance in the organic matter.

Surprisingly little attention has been given to the amount of roots different crops produce. Recent research in organic matter management suggests that it is mainly the roots that contribute to the formation of active organic matter. Crops with considerable amount of roots, like perennial forage, are known to be real soil builders. The root mass formed by these crops can be very high. Not only the quantity of residues is important but also the quality.

The chemical composition determines whether they contribute to the formation of organic matter.

Nitrogen, carbon and lignin have been shown to be substances that determine to a large extent the value of residues in the soil. Carbon rich materials like straw (C/N ratio of 40 to 100) tend to initially break down slower than nitrogen rich materials like alfalfa or vegetable residues with a much lower C to N ratio (5 to 20). Microorganisms need, however, a certain amount of nitrogen and preferably a C/N ratio around 10 to multiply. As carbon rich materials decompose, these organisms are likely to tie up the nitrogen available in the soil. Decomposition of carbon rich materials like straw can therefore result in a shortage of nitrogen for plant uptake. In addition, a less efficient microbial metabolism can occur during the decomposition, which results in a larger loss of carbon as CO_2 during respiration. This leaves less material to form quality organic matter.

Cover Crops and Green Manures

Cover crops and green manures generally contribute to soil organic matter formation. Their contribution to building the soil greatly depends, however, on the age of these crops. If seeded into a row crop or sequentially seeded after the main crop they can contribute to the quality of the soil simply by keeping the soil covered and thus limiting erosion. This planted canopy will prevent soil structure loss by breaking the force of raindrops in rainstorms. The roots penetrating the soil will enhance soil life and structure. In the fall, the cover crops continue to take up nutrients and thereby prevent leaching from the system.

Table 3. Amount of residues from different crops.

Crop	*Crop residue (lb/a)*
Corn	6100 - 9100
Soybean	2500 - 5000

Wheat	2400 - 4500
Oats	1600 - 2400
Cover crop (clover)	900 - 4900
Cover crop (oats, rye)	1000 - 5500

A cover crop worked into the soil will act as a green manure by providing the following crop with essential nutrients. The length of the growing season of the green manure and its maturity will determine its nutrient contribution to the next crop and its effect in building soil organic matter. During the growth of a green manure the biomass will increase but the nitrogen concentration of tops and roots will generally decrease.

Young material with a higher N concentration (low C/N ratio) will therefore provide nitrogen for the following crop but contribute little to the organic matter formation due to the limited biomass and low C/N ratio of the residues. If the green manure is left to grow in the field for a longer period, it will form more biomass. Worked into the soil this more mature material will do a much better job in forming organic matter, but little nitrogen may become available in the short run for the following crop. A balance needs to be created between these soil building practices and destructive practices such as tillage. Perennial legumes, grasses, tillage practices and regular manure applications are really important for creating a healthy balance and creating a healthy soil over time.

Compost and Manure

Manure and compost can make a strong direct contribution to building soil. The composting of manure can be managed to produce young organic matter with a high value for soils. Fresh manure and young, rich composts may be a valuable source of nutrients (especially nitrogen) for crops. They may also have strong effects on stimulating the formation of crumbs in the soil. Older, finished composts are more stabile and release less nitrogen to crops. They build active organic matter in the soil and stimulate root growth.

During the composting pro-cess a kind of digestion takes place. This digestion consists of decomposition, transformation and curing (maturing) processes. Organic matter and nutrients, air, moisture and living organisms are the most important ingredients for a successful composting process. The materials used, the method of composting and temperature during the process determine the final product. Fully mature compost looks like rich soil. The plant residues are no longer recognisable as such.

Table 4 indicates the preferred conditions for a good composting process. Best results can be obtained if original materials are used with a C/N ratio of about 25 to 40:1. A higher C/N ratio than this, caused by adding large quantities of sawdust or straw, results in a slow composting process in which large losses of CO2 occur due to a less efficient process at the microbial level. As compost matures the C/N ration will decrease to 10 to 15:1 Materials with a low C to N ratio result in a quick transformation with high temperatures in the compost piles in which losses of nitrogen to the atmosphere as ammonia are very likely.

The result might be a product with a high nitrogen supplying capacity to crops but less ability to build soil (Table 5). It is not the purpose to discuss composting process techniques and methods here in detail. Many good books and papers are available on this subject. However, the final product can have pronounced effects on the organic matter formation in the soils.

Table 4. Recommended conditions for composting

Condition	*Range*
Carbon to nitrogen (C/N) ratio	25 to 35:1
Moisture content	50 - 60 %
Oxygen concentrations	> 5%

Particle size (inches)	1/8 - 1/2
pH	6.5 - 8.0
Temperature (oF)	130 - 140

Maturity of the compost influences how it performs in the soil (Table 6). Relatively young compost that is 2 to 3 months old is likely to have a more direct plant feeding capacity. If the C/N ratio of the product is high (>20), a serious tie-up of nutrients might occur in the soil. Materials with a low C/N ratio release N more quickly. This means that nitrate and ammonium are directly available for plant uptake, but are also easily lost through leaching and volatilisation. Such materials are suited for some vegetable crops, corn and wheat. More mature compost that has been composted for about 12 months is more likely to act as a soil builder. The contribution to the active fraction of organic matter is likely to be higher. Nutrients become available more slowly and nutrient losses are less likely.

Table 5. Quality characteristics of a selection of composted and manure residues

Manure	*% moisture*	*C/N*	*% N*	*lignin %*
Cattle slurry	81	15	2.4	20
Horse	72	29	1.6	14
Compost (cattle)	75	17	2	26
Sheep	69	16	2.7	21
Swine	80	14	3.1	na
Laying hens	69	6	8	6
Vegetables	87	19	2.7	6
Leaves	38	54	0.9	na

ORGANIC MATTER BUDGETING

Organic matter budgeting is meant as a tool for long-term soil management. With this budgeter, estimates are made of how crop rotations or manure applications affect the amount of active organic matter that residues and manures contribute to soil over time. Estimates are also made as to how much nitrogen becomes available in the soil for plant uptake. This system is meant to provide a planning tool for farmers to design sustainable farming systems and to help them gain insight into the processes that take place in their soils. The budgeter is not a finished product, but is constantly being refined, improved and adapted to fit different regions. We are dong this work through a series of grant funded, on-farm research projects in the Upper Midwest.

The budgeter has been developed as an Excel spreadsheet. Budgeting can also be done by hand and this bulletin shows how to use the budgeter through examples of different crop rotations and manuring programmes. Tables and worksheets are provided so readers can work out organic matter budgets for their individual rotations and farming situations. Simple calculations can be done manually, or with a pocket calculator.

Table 6. Working of compost and manure

	composted straw manure	*straw manure*	*fresh manure*
C/N ratio	16-24	24-30	6-13
nutrient uptake	+	+	+++
working	-	+	+++
losses	++	+	--
disease control	+++	+	-+
water holding capacity	+++	+	-
humus formation	+++	+	-

The budgeting system's advantage is that it can be used to evaluate the impact of various crop rotations, animal manures and green manures, and thereby serve as a cost-saving, ecologically sound planning tool. If the active organic matter increases over a crop rotation cycle this probably indicates a sound crop rotation and increasing soil quality. A decline in the amount of active organic matter means that changes may be necessary to prevent a decline in soil quality.

Factors such as climatic conditions, soil type, and history of the land, crop rotation, and management practices like manuring and tillage determine the amount of active organic matter found in the soil. After native soils were put into agricultural use in the Midwest there was a considerable decline in the amount of active organic matter. It is important to find out how to maintain or increase this active fraction of organic matter in a particular cropping system's soils. However, increasing the amount of soil organic matter is difficult and takes several years. Increases in the overall soil organic matter level are generally limited to about 30 percent under highly favourable conditions. However, if soil is put into extensively managed pasturage or high levels of manures and composts are applied, organic matter increases may surpass 30 percent.

In a farming system, several sources of organic material can serve to produce the active organic matter. Plant residues from roots and straw determine to a large extent the annual input. In a sustainable farming system, legumes grown as green manures or animal manures can contribute to the input. To do the budgeting we need to account for the crop rotation, the amount of residues produced by crops and the composition of these residues. Within a rotation, fallow years or prolonged periods without soil cover will result in erosion and a loss of the active organic matter.

A forage based system, with soils kept under grasses or grass-legume mixtures will result in general in an increasing active organic matter fraction. Soil tests are necessary to determine the active organic matter for a particular field. Table 7 gives us an example of a crop table used in the budgeting system. The table summarises the nitrogen release in the soil from corn residues and their contribution to the fraction of active organic matter depending on the yield level. In the budgeting we accounted for 2 years of nitrogen release (a minus sign indicates a nitrogen tie-up) from the residues following the harvest.

During decomposition, residues are transformed partly into active organic matter. In the third year after crop harvest, residues will contribute to the active organic matter fraction. Straw or stover is a source for forming active organic matter if it is left in the field. In the table we distinguish between the two options "straw removed from the field" and "straw left in the field." Depending on the specific farming practices and markets, farmers keep their straw in the field, or remove it to use as bedding or to sell off the farm. If a crop is removed from the field for silage or haylage we use a table yield estimate for the crop based on what we think it would have yielded had we grown the crop out and harvested it for grain.

Table 7. Nitrogen release (lbs N/acre) and contribution to the active organic matter (lbs C/acre) of corn residues for different yield levels.

stover removed

Grain yield. bu/acre	*N release year 1*	*year 2*	*C to the active organic matter*
60	-10	2	250
70	-11	3	291
80	-12	3	333
90	-12	4	374
100	-13	4	416
110	-13	5	458
120	-14	5	499
130	-15	6	541
140	-15	6	582
150	-16	7	624
160	-17	7	666

170	-17	8	707
180	-18	8	749
		stover left	
	N release		*C to the active*
	year 1	*year 2*	*organic matter*
	-21	5	405
	-22	5	462
	-23	4	520
	-24	4	578
	-25	4	636
	-26	3	694
	-27	3	751
	-28	3	809
	-30	2	867
	-31	2	925
	-32	2	983
	-33	1	1040

Legumes contribute to the active organic matter. They are essential in sustainable farming systems because they provide the main source of external nitrogen due to their capacity to fix atmospheric nitrogen in root nodules. If used as a cover crop, legumes may also limit soil losses by erosion and also limit the rapid organic matter turnover found in bare soil.

These tables are sufficient for estimating the contribution of forages during the establishment year and the first year of production. After that, the contribution of the perennial legumes is hard to estimate. While they are growing and are being harvested, large quantities of roots grow and decay and large quantities of leaves are lost from the hay and decompose in the soil. In fact several studies indicate that while alfalfa is being cropped, large quantities of roots, equal to the weight of roots in the spring, are turned over during the growing season. This makes it important to estimate how many roots alfalfa really has. Alfalfa tends to produce a maximum quantity of roots in its second or third year of growth. After that, many plants disappear from the stand due to disease, but the remaining plants grow larger roots to compensate. This compensation is not complete, and gradually, as the stand ages, the quantity of alfalfa roots in the soil declines.

Grass roots grow slowly, but after several years of production, the quantity of grass roots may greatly exceed that of alfalfa, though the alfalfa may still produce many more tops than the grasses. Mature grass plants have a low top/root ratio and they can produce massive quantities of roots. These grass roots are excellent at building soil structure and packing organic matter into soil crumbs. The grass plants are more competitive than alfalfa at taking up nitrogen from decaying soil organic matter. They force the individual alfalfa plants to fix a greater percent of the nitrogen than they would if the alfalfa were growing alone.

Relationships between hay production and carbon turnover are also complex. Because of this, we have developed a computer submodel in the spreadsheet budgeter to estimate the C turnover and N released by the crop. It is important to know the yield of the alfalfa stand, the percent alfalfa in the hay, the age of the stand, and the percent protein in the hay. This can be easily done with the spreadsheet version of this organic matter budgeter, but it is more difficult to do accurately by hand. Therefore, we refer the reader to the spreadsheet in order to make these calculations.

Manures and compost are added to a cropping system to improve soil quality and to serve as a source of nutrients for crops. Their contribution to active organic matter can be considerable, depending on the composition and stage of decomposition. The storage method, type of manure, and quantity of added straw have a large effect on the decomposition of the manure in the soil. We are aware that the nitrogen release from manure applications that we have predicted here is generally lower than common recommendations. However, we expect that some of the benefits to yields that are associated with manure applications are not due to nitrogen release, but rather to improved root health and root access to nutrients.

Table 8. Estimated size of the active organic matter fraction for a Warsaw silt loam

Organic matter %	*Active organic matter) (lbs C/acre)*	*Organic matter %*	*Active organic matter) (lbs C/acre)*
2.0	13819	3.6	24874
2.4	16582	4.0	27637
2.8	19346	4.4	30401
3.2	22110	4.8	33165

Table 8 summarises the correlation between the total organic matter content of a soil and the amount of active organic matter for a Warsaw silt loam commonly found in southeastern Wisconsin. We determined this correlation by running the Century simulation model for a silt loam based on past land use. We will use these values for the active organic matter in the examples of the organic matter budgeting system. It should be realised, however, that the correlation could vary. Intensively cultivated soils are likely to have less active organic matter.

Soils that have been in pasture for prolonged periods tend to have higher levels of active organic matter. Lighter soils have less active organic matter; heavier, clayey soils tend toward more active organic matter. Therefore, these values are only approximations. In the research work with farmers using this budgeter we are estimating the active organic matter by multiplying the particulate organic matter content (POM) times 2. Particulate organic matter is organic matter about the size of sand, and it appears to make up half of the active organic matter in the soils. Loss of active organic matter is mainly determined by its rate of decomposition and by soil erosion. The type of crops, climatic factors and tillage practices all influence the rate of decomposition.

Table 9. Amount of active organic matter lost each year under different crops.

Crop	*Decomposition factor (-)*
Corn	0.0469
Soybean	0.0353
Winter wheat	0.0469
Barley	0.0533
Oats	0.0554
Potatoes	0.0595
Sugar beets	0.0566
Red clover	0.0164
Alfalfa	0.0218
Sweet clover	0.0218

Variation can occur due to local climatic conditions and specific farming system practices. Table 9 summarises the fraction of active organic matter lost annually under different crops grown on a silt loam in southeastern Wisconsin. These losses take into account climatic cond-itions, general management practices for the area, and soil erosion.

REFERENCES

Ahern, C., Weinand, M., and Murtha, G., "Surface soil pH in non-arid Queensland", *Aust. J. Soil Water Conserv.* 5, 38-43, 1992.

Bell, M.J., Moody, P.W., Connolly, R.D., and Bridge, B.J., "The role of active fractions of soil organic matter in physical and chemical fertility of Ferrosols", *Aust. J. Soil Res.* 36, 809-19, 1998.

Dalal, R.C., and Mayer, R.J., "Long-term trends in fertility of soils under continuous cultivation and cereal cropping in southern Queensland", VII. Dynamics of nitrogen mineralisation potentials and microbial biomass, *Aust. J. Soil Res.* 25, 461-72, 1987.

6

Biofertilisers and Sustainable Farming

The name itself is self explanatory. The fertilisers are used to improve the fertility of the land using biological wastes, hence the term biofertilisers, and biological wastes do not contain any chemicals which are detrimental to the living soil. They are extremely beneficial in enriching the soil with those microorganisms, which produce organic nutrients for the soil and help combat diseases. The farm produce does not contain traces of hazardous and poisonous materials. Thus those products are accepted across the world as Organic ones. Hence for organic farming the use of biofertilisers is mandatory.

Biofertiliser is a large population of a specific or a group of beneficial microorganisms for enhancing the productivity of soil either by fixing atmospheric nitrogen or by solubilising soil phosphorus or by stimulating plant growth through synthesis of growth promoting substances. Bio-fertilisers based on renewable energy source are cost effective supplement to chemical fertilisers, eco-friendly and can help to economise on the high investment needed for chemical fertiliser use as far as nitrogen and phosphorus are concerned.

Biofertiliser is a 100% natural and organic fertiliser that helps to provide and keep in the soil all the nutrients and microorganisms required for the benefits of the plants. There are millions of Microscopic Organisms near the plants that conform a in a micro environment that provides nutrients to the plants also helps to keep the water and retain the nutrients in the soil easy available to the plants. When you fill them with chemicals, most of them die forever, losing the capacity of the soil to be sustainable at long term.

The end product of the various phases of Biodigestion, is a liquid which is dark in colour due to the presence of Humus in it. We call this Pure Biofertiliser, and it can be used on soil as a high quality organic fertiliser, or even as a corrector of pH, bacterial life and texture. The Biofertiliser has a relative high nutrient concentration, and even so, it can be used directly over soil before planting. Once diluted, it constitutes a high quality foliar fertiliser, and in this form, it is known as Diluted Biofertiliser.

Diluted Biofertiliser also has all the needed conditions to be used as a complete Nutrient Solution in Organic Hydroponics. The advantages of using the biofertiliser are enormous, not only it is very economical, but also for the high agricultural yields which it produces.

USE OF SOIL MICROORGANISMS

Most fertilisers add nitrogen to the soil. This can be done via chemical fertilisers, or through a process called biological nitrogen fixation (BNF). On a worldwide basis it is estimated that about 175 million tons of nitrogen per year is added to soil through biological nitrogen fixation (BNF). The term bio means *living*; so bio-fertilisers refer to living, microbial inoculants that are added to the soil. These bio-fertilisers are products consisting of selected and beneficial microorganisms, which are known to improve plant growth through supply of plant nutrients.

The soil microorganisms used in biofertilisers are: Phosphate Solubilising microbes, Mycorrhizae, Azospirillum, Azotobacter, Rhizobium, Sesbania, Blue Green Algae, and Azolla. Let's go through these groups in a little detail in order to understand their role in biofertilisers, which can be used to make rich, living soil that is suitable for coffee plants.

Phosphate Solubilising Microbes

Phosphorus is an important nutrient for plants. There are several microorganisms which can solubilise the cheaper sources of phosphorus, such as rock phosphate. Bacteria like Pseudomonas striata, and Bacillus megaterium are also important phosphorus solubilising soil microorganisms. Many fungi like Aspergillus and Penicillium are potential solubilisers of bound phosphates. They solubilise the bound phosphorus and make it available to the plant, resulting in improved growth and yield of crops.

Soil phosphates are rendered available to plants by soil microorganisms through secretion of organic acids. Therefore, phosphate dissolving soil microorganisms play some part in correcting phosphorus deficiency in plantation soils. They may also release soluble inorganic phosphate into soil through decomposition of phosphate rich organic compounds. These microbial inoculants can substitute almost 20 to 25% of the phosphorus requirement of plants. Phosphate solubilising microbes can also be inoculated to coffee husk along with rock phosphate while preparing compost to enrich the compost with available phosphorus.

Mycorrhizae

The term "mycorrhizae" refers to fungus associated with plant roots. These fertilisers are divided into ectotrophic and endotrophic or the vesicular arbuscular mycorrhiza (VAM) categories. Most plants depend on their mycorrhizal association for adequate uptake of nutrients (especially the immobile ions such as phosphate, zinc and micronutrients) and survival in natural ecosystems. Mycorrhixal association stimulates branching of the root and increases the absorption surface of the root. Other benefits include tolerance to drought, high soil temperature, soil toxins, and extreme Ph levels, as well as protection against root pathogens. This is why, When trees are introduced to new regions, inoculation of soil with mycorrhizal fungi is a necessary prerequisite for the establishment of the trees.

Azopirillum

Azospirillum are nitrogen-fixing bacteria that lives in a symbiotic relationship in the root cortex of several tropical crops. They stimulate plant growth through nitrogen fixation and production of growth substances like auxins, gibberellins and cytokinin. It is estimated that almost 10 to 15% of the required nitrogen can be met by Azospirillum biofertiliser.

Azobacter

Azotobacter are free-living, nitrogen-fixing bacteria and are known to produce several plant growth promoting substances. In addition to nitrogen fixation by these bacteria, they are also known to protect plants against pathogenic microorganisms either by discouraging their growth or by destroying them. These inoculants need more attention in view of their triple action of nitrogen fixation, bio-control, and production of plant growth regulators.

Rhizobium

Rhizobium bacteria, basically form root nodules in leguminous plants and fix atmospheric nitrogen in a symbiotic association. The Rhizobium bacteria gives nitrogen to the plant and the plant gives protection to the bacteria from oxygen damage by harbouring it inside the root nodule.

Sesbania

Many legumes are grown and then turned into the soil

while they are still green to enrich soil nitrogen. Sesbania is a green manure plant which forms both root and stem nodules in association with rhizobium and thereby fixes more atmospheric nitrogen. These legumes produce ten times more nodules than other legumes and have a very high capacity to fix atmospheric nitrogen. Apart from enrichment of soil nitrogen, green manuring enriches the phosphorus, calcium, sulphur and other micronutrient of the soil.

Blue Green Algae

Blue Green Algae (BGA) or Cyanobacteria have the ability to carry out both photosynthesis as well as nitrogen fixation. They belong to the order Nostocales and Stigonematales. Algal flakes are grown and then broadcasted.

Azolla

Azolla is a floating fern which harbours a blue green algae in its leaf cavities. The fern multiplies very fast with the symbiotic association of the algae and this rapid multiplication creates a huge amount of biomass on the surface of the water. It is then harvested, dryed and used as biofertiliser.

BIOFERTILISERS FOR SUSTAINABLE AGRICULTURE

For a sustainable agriculture system, it is imperative to utilise renewable inputs which can maximise the ecological benefits and minimise the environmental hazards. One possible way of achieving this is to decrease dependence on use of chemical nitrogen fertilisers by harvesting the atmospheric nitrogen through biological processes. With these considerations, the Department is supporting a programme on R&D and application of biofertilisers. A network project on Development of Transgenic Biofertilisers which are efficient Nitrogen fixers and

Phosphate solublisers, was launched in the 10th plan period. Under this programme development of transgenic biofertilisers, has been undertaken by increasing copy number of *nif* genes; incorporation of *hup* genes wherever *hydrogenase* enzyme is not produced and knocking off negative regulatory gene from the regulatory operon.

Achievements

Under Integrated Nutrient Management (INM) studies, biofertiliser component has not received adequate attention in the past. Keeping this in view, a National Network Project was launched in 1999 for 3 years on the 'Role of biofertilisers in INM' embracing 17 centres throughout the country. The objective was to establish useful biofertiliser packages for yield optimisation of some major crops in various cropping systems in different regions. In addition to generating field data on biofertilisers used in integration with reduced levels of chemical fertilisers, advanced molecular tools were employed for genotypic characterisation of microbial strains for their identification and performance evaluation in the field.

A number of INM packages have been developed for Poplar-Eucalyptus, Mungbean-Wheat-Bajra, Rice-wheat, Rice-Rice, Sugarcane, Rice-Wheat-Pulse and saline area based cropping systems. The packages developed were demonstrated to about 5000 farmers.

Leads Obtained

Phosphoenol pyruvate carboxylase (*pqq*) synthase gene has been transferred to *Azospirillum*. It was found that the transconjugant *Azospirillum* now possess mineral phosphate solubilising (MPS) ability. No change was observed in the nitrogen fixation of these transconjugants. The *pqq* synthase gene was isolated from *Burkholderia cepacia* via PCR amplification of primers designed using the conserved sequence across the gene reported in different species and cloned into *E coli*. The *E coli* containing the *pqq* synthase gene now showed MPS activity. These results indicate the possibility of developing transgenic bacteria by cloning pqq synthase genes into certain gram-negative rhizobacteria. The gene has also been electroporated into *Rhizobium* and *Azotobacter* and there it has shown MPS activity.

Glucose dehydrogenase gene (*gdh*) of *Bacillus licheniformis* has been successfully amplified using gene-specific primers. The amplicon (802bp) was then cloned into pTZ57. The clone was confirmed as *gdh* gene by sequencing. Primers have been designed for the *gdh* sequences of *Serratia marcescens* and *Pseodomonas aeruginosa* and the genes would be amplified using specific primers. Pure Genomic DNA has been isolated and attempts are being made to construct a library in pUC18. The library would be screened for the *gdh* gene clone based on the zone of solubilisation. Selected recombinants would be probed for the presence of the gene using heterologous *gdh* probes.

Technologies

One of the major concerns in today's world is the pollution and contamination of soil. The use of chemical fertilisers and pesticides has caused tremendous harm to the environment. An answer to this is the biofertiliser, an environmentally friendly fertiliser now used in most countries. Biofertilisers are organisms that enrich the nutrient quality of soil. The main sources of biofertilisers are bacteria, fungi, and cynobacteria (blue-green algae). The most striking relationship that these have with plants is symbiosis, in which the partners derive benefits from each other.

Plants have a number of relationships with fungi, bacteria, and algae, the most common of which are with mycorrhiza, rhizobium, and cyanophyceae. These are known to deliver a number of benefits including plant nutrition, disease resistance, and tolerance to adverse soil and climatic conditions. These techniques have proved to be successful biofertilisers that form a health relationship with the roots. Biofertilisers will help solve such problems as increased salinity of the soil and chemical run-offs from the agricultural fields. Thus, biofertilisers are important if we are to ensure a healthy future for the generations to come.

NUTRIENT MANAGEMENT IN ORGANIC PRODUCTION

Biofertilisers are materials of biological origin, which are added to deliver the nutrients contained in them such as green manure and organic inputs. Organic nutrient management was evolved for various crops for overcoming the adverse effect of chemical inputs and to sustain crop productivity. Biofertilisers are carrier-based inoculants containing cells of efficient strains of living organism like bacteria (Cyanobacteria), algae (Blue Green Algae) and fungi (Mycorrhizae) which are capable of fixing atmospheric nitrogen or solubilising soil phosphorus or stimulating plant growth. Biofertilisers are cost effective, eco-friendly and renewable energy source of plant nutrients.

Microorganisms that play crucial role in the fertility management are eighter free living or having symbiotic association with plants. Based on the type of microorganisms, biofertilisers are classified as follows:

1. Bacterial biofertilisers: eg. Rhizobium, Azospirillum, Azotobactor, Phosphobacteria
2. Fungal biofertilisers: eg. Mycorrhiza
3. Algal Biofertilisers: eg. Blue Green Algae (BGA) and Azolla, and Actinomycetes biofertilsers : eg. Frankia.

Broadly the biofertilisers are also classified into two types: a) biological nitrogen fixing biofertilisers and b) the phosphate solubilsing or mobilising biofertilisers. Nitrogen fixing biofertilisers consist of microorganisms which fix atmospheric nitrogen eighter symbiotically or asymbiotically into the plants, whereas, phosphate solubilising biofertilisers have the ability to solubilise or mobilise the fixed insoluble phosphate soil.

Rhizobium Biofertilisers

Most of the cultivated tropical and subtropical soils have good population of index of establishment of an efficient nitrogen fixing system. Competitive and efficient strains of Rhizobium used for inoculation will ensure maximum nitrogen fixation. Absence of effective Rhizobium strains is responsible for poor nodulation encountered in many areas. Absence of suitable Rhizobia, deficiency or toxicity of a nutrient, unfavourable soil conditions, excess water logging, unsuitable pH, predators, pests are other factors which indirectly influence the potentiality of Rhizobium strains.

Drought and temperature stress is known to affect the growth of legume host and nitrogen fixation. Nodule formation and nirtogen fixation are influenced by the rhizosphere temperature. The capacity to tolerate elevated temperature very considerably among different strains of rhizobia and host species. Therefore, the ability to tolerate higher temperature will be a very desirable trait for selection of rhizobium'strains in a specific location as this would help in build up and carryover of strains introduced into soil from one season to another by tidying over higher soil temperature during summers.

The *Rhizobium* strains having significant potential of nitrogen fixation ranging from 50-300 kg/N/ha per season depending on the crops. Further legumes often require specific strains of *Rhizobium* for maximum nitrogen fixation and nodule formation. On the basis of specificits, the genus *Rhizobium* consists of different groups named as *Rhizobium leguminoserum* which further have three biovars *(trifolii, phaseli* and *viceae), Rhizobium meliloti, Rhizobium loti, Rhizobium fredii, Rhizobium ciceri, Rhizobium jampnicum,* and *Rhizobium spp.* The main effective strains of *Rhizobium* are listed in the table appended

Azotobacter Biofertilisers

Azotobacter is a heterotroph bacterium of aerobic character having the capability of fixation of dinitrogen (N2) as nonsymbiont. Azotobacter not only provides the nitrogen but also produce a variety of growth promoting substances. Some of them are indole acetic acid (1M), gibberellins (GA), Vitamin-B, and antifungal substances. Another important characteristics of Azotobacter associated with crop improvement is excretion of ammonia in the rhizosphere in presence of root exudates.

The benefits of Azotobacter inoculation on crop production are well known. barley, maize, paddy, jowar, oat, sugarcane, tobacco, cotton, potato, brinjal, onion, cauliflower, cabbage, other vegetables, sunflower mustard, seasemum, linseed, tea, coffee and all type of forest fruit and flower plants are known to fix 20-30 kg. N/ha. Azotobacteralso causes mineralisation of fixed phopphate of soil and thus increase uptake of phosphate in plants. The identified effective strains of Azotobacterare listed in the table appended.

Azospirillum Biofertilisers

Azospirillum an associative micro-aerophilic nitrogen fixer, is commonly found in association with the roots of cereals and forage grasses, High nitrogen fixation capacity, low energy requirement and abundant establishment in the roots of cereals and tolerance to high soil temperature (30-40 C) are responsible for itssuitability under tropical conditions too. Azospirillum is also found on root surface, root hairs, epidermal cells and vesicular tissues.

Azospirillum are mesophilic and reported in association of crops grown in acidic to alkaline pH range. Azospirillum are metabolically versatile and can grow vigorously in the presence of nitrogenous compounds present in soil but as soon the external combined nitrogen supply is exhausted the bacteria switch on to diazotrophy.

Azospirillum inoculant consistently help in increasing grain yield of paddy, barley, bajra, ragi, sugarcane and sorghum. It is also beneficial for vegetables, fodder crops, fruits and flower plants. Azospirillum inoculation is found to fix 20-30 kg.N/ha per crop season and maintain soil fertility in semi-tropic and tropic regions. The effective and tested strains of Azospirillum are listed in table appended.

Phosphate Solubilising Biofertilisers

The efficiency of utilisation of phosphatic fertilisers is very low due to chemical fixation in soil, Besides, native soil phosphorus is mostly unavailable to crops because of low solubility. Indian soils are characterised by poor to medium status with respect to available phosphorus. Further, there is building of insoluble phosphates in soil where phosphatic fertilisers have been applied over long periods. The introduction of efficient phosphate solubilisers in the rhizosphere of crops and soils increases the availability of phosphorus from insoluble sources of phosphates and utilisation of efficiency of phosphatic fertilisers such as superphosphates.

Important phosphate solubilising organisms are Pseudomonas straiata Bacillus polymyxa) Aspergillus awamori, Penicillium digitatum. These micro-organism can grow on insoluble phosphatic sources such as tricalcuim phosphate, ferric, aluminium and magnesium phosphate, rock are implicated in the solubilisation and mineralisation of insoluble and organic phosphate produced has important bearing on amount of phosphate solubilised. The important strains of phosphate solubilising biofertilisers are listed in table appended.

Techniques of Application of Rhizobium Biofertilisers

Treatment of seeds with the slurry of Rhizobium inoculant is the right and effective method of application for getting maximum response. The method of application is as follows:

i) Prepare the slurry of required quantity of inoculant in sufficient water.

ii) Pour this slurry over the heap of seeds to be treated.

iii) Mix the seeds homogeneously over dean floor or on plastic sheet or on gun bag for drying purpose.

iv) Sow the treated seeds immediately.

If phosphate solubilising biofertiliser (phosphatical) is also available for treatment, suspend equal quantity of Rhizobium and Phosphatical (1: 1 ratio) in 400-500 ml water and mix thoroughly, spread in shade for drying and sow immediately.

Doses: 10 Kg of normal size seeds such as urdbean, pigeon pea, cowpea, ricebean, vegetables like mehti and other such legume seeds may be treated with 200 gm of Rhizobium inoculant by slurry method. The big size seeds such as sobean, gardenpea, french beans, rajmash, wingbean, etc.

Techniques of Application of Azotobacter Biofertilizers

Azotobacter is a broad spectrum inoculant useful for so many crops. The inoculant maybe used in different ways depending upon the crop to be cultivated. the different recommended methods of application etc.

a) Seed treatment;

b) Seedling treatment;

c) Sets/tuber treatment; and

d) Soil treatment.

Seed treatment

The method of seed treatment is same as described in the case of Rhizobium. If phosphate solubilising bacteria (Phosphatica) is to be treated simultaneously, the same method may be used (1:1 ratio) as described in the case of Rhizobium.

Doses: 10 kg medium size seeds such as maize, wheat, barley, cotton, cucurbit vegetable seeds etc. may be treated with 200 gm Azotobacter inoculant, Whereas, 100 gm of inoculant is enough for treatment of very small size seeds such as mustard, finger millet, cabbage, cauliflower and other cole crops, tomato, brinjal chilli, capsicum and tree tomato, radish, carrot, turnip, palak, amaranthus, leafy mustered, onion seeds, etc.

Seedling treatment

This method is used where the transplantation of seedlings are required. The method of application is as follows:

i) Prepare the suspension of required amount of inoclant in water in the ratio of 1:10) or 1 :1:1 0 ratio) of Azotobacter: Phosphatica : Water).
ii) Dip the roots of seedlings in suspension and keep them immersed for about five minutes.

Take out the seedlings from the suspension and transplant them immediately. This method is effective for paddy finger millet, vegetables such as cabbage, cauliflower, broccoli, tomato, brinjal, chilli, capsicum, onion, True Potato Seedling fruit seedlings and grafts/buildings and other seedlings treatments.

Doses: Suspension of 2.5 kg each of Azotobacter and Phosphatica in 25 to 40 litres of water sufficient for treatment of seedlings enough for one ha.

Sets/tuber treatment

This type of application is beneficial for sets of sugarcane, tubers of potato, Ginger rhizome, Turneric rhizome, Garlic cloves, sets of cassava, cuttings of sweet potato, fruit seedlings/·grafts/buildings and flower plants.

i) Prepare suspension of one Kg each of Azotobacter and Phosphatica inoculant in 40 to 60 liters of water.

ii) Immerse the sets/tubers in the suspension for 5 to 10 minutes.

iii) Bring out the treated sets/tubers and plant them without any delay.

This method is time consuming and requires a lot of labour. However, the results of the method are very significant. An alternate spray method may be used which is item and labour saving. The spray method is described as follows:

i) Prepare the suspension of required amount of inoculant and filter this with sieve or cloth.

ii) Take the filtrate in a clan spray machine and spray it over the heap of sets/tubers.

iii) Mix up the sets/tubers during spray so that all the sets/tubers get treated.

iv) Sow the treated sets/tubers immediately.

Doses: Suspension of 2.5 kg each of Azotobacter and Phosphatica inoculant in 80 to 120 litre of water is required for treatment of sets/tubers enough for one ha of land.

Soil treatment

Prepare the mixture of required amount of inoculant in top soil/FYM and broadcast this mixture at the time of last ploughing or at the time of irrigation in standing crops.

Doses: 5 to 8 kg each of Azotobacter and Phosphatica inoculant mixed with 100 to 120 kg soil/FYM/Compost is the requirement for short duration crops (less than six months) for one hectare of land. Double this dose for long duration (perennial) crops.

Azospirillum and Phosphate Solubilising Biofertiliser

The doses and methods of application for Azospirillum and phosphate solubilising biofertilisers are same as described in Azotobacter inoculant.

Precautions

1. Always treat the seeds in shade;
2. Do not expose the treated seeds to direct sun rays;
3. Use only specific Rhizobium meant for specific legume;
4. Treat the seeds immediately or just two or three hours before sowing;
5. Do not mix any chemical fertilisers and pesticides directly with inoculant;
6. Store biofertilisers in cool and dry place away from heat and direct light;
7. Open the packet just before use and use all its contents at a time;
8. Use double.quantity of biofertiliser on seeds treated with mercurial compound;
9. Use biofertiHzers before expiry date specified on each packet.

Transfer of technology and launching of product

Mycorrhiza colonise 85% of land plants, offers 50 percent saving in chemical (phosphorus) fertilizer, higher resistance to soil-and-root borne pathogens, and helps in wasteland reclamation by providing extended arm to plant root system. Under the DBT supported project TERI, New Delhi developed technology package for Mass production of suitable Mycorrhisa.. The technology has been transformed to two industries namely M/s Cadila

Pharamaceutical Ltd., Ahemdabad, and KCP Sugar and Industries (Pvt.) Ltd., Chennai. These industries have already started manufacturing the mycorrhiza based biofertiliser and launched it in the market.

The product is in great demand and KCP Sugar and Industries (Pvt.) Ltd., Chennai, which has launched the products under the brand name "Ecorrhiza-VAM' and 'Nurserrhiza-VAM' etc has been awarded by the All India Biotechnology Association. M/s Cadila Pharmaceuticals, Ahmedabad has also launched the product in the name of 'Josh – a root booster' and it is being used by the Gujrat State Forest Department. Mass production technology of Rhizobial biofertilisers transferred to two entrepreneurs namly; M/S Javery Agro, Amrawati and M/S Pratistha Industries Ltd., Secunderabad. However, the product is yet to be marketed.

Initiatives

A new network programme associating 11 centres launched for the development of transgenic biofertilisers with better nitrogen fixing and phosphate solubilising ability. In addition efforts are being made to generate some more projects on the following aspects: Generating applied and adoptive research programme for the development of biofertiliser based nutrient management practices of high value plantation crops, spices, medicinal and aromatic plants, protected cultivation and organic farming of high value horticultural crops etc.

Entrepreneurship development among public and private sector institutions, organisations etc. for small scale production and delivery of biofertiliser inoculants with attention to quality control, farmers' advisory and adoption. Developing technology packages for integrated plant health and production management of crop plants through the use of diverse bioinoculants in combination. Using molecular transgenosis and genetic engineering techniques to improve upon or enhance the primary functional attributes of biofertiliser microorganisms, particularly the nitrogen fixing ones, and to solve the critical problems of their ecological adaptation to and adjustment in the rhizosphere of soil-plant systems; Designing biofertiliser microorganisms with multidimensional attributes.

Creating data base for defining the protocol and laying out the strategies for releasing the transgenic biofertiliser strains in the environment keeping environmental safety into consideration; Programmes designed to analyse and assess the impacts of biofertiliser use in production systems for fine tuning of biofertiliser technology.

Biofertilisers are ready to use live formulates of such beneficial microorganism which on application to seed, root or soil mobilise the availability of nutrients by their biological activity in particular and help build up the micro flora and in turn the soil health in general. Nitrogenous biofertilisers harvest atmosphere nitrogen and converts into ammonical form, which in due course is made available to the plants or is released in the soil. Phosphate solutions solubilise fixed forms of phosphorus already present in the soil and make it available for use of plants. Compositing biofertilisers are used for hastening the process of compositing and for enriching its nutrient value.

With the introduction of green revolution technologies the modern agriculture is getting more and more dependent upon the steady supply of synthetic inputs (mainly fertilisers) which are products of fossil fuel (coal+ petroleum). Excessive dependence of modern agriculture and the supply of these synthetic inputs and the adverse effects being noticed due to their excessive and imbalanced use has compelled the scientific fraternity to look for alternatives. Some of the points of concern are:

i) Availability and cost:

a) Demand is much higher then the availability. It is estimated that by 2020, to achieve the targeted production of 321 million tones of food grain, the requirement of nutrient will be 28.8 million tones, while their availability will be only 21.6 million tones being a deficit of about 7.2 million tones.

b) Increasing costs are getting unaffordable by small and marginal farmers.

ii) Effect of Chemical fertilisers in soil and environment.

a) Excessive and imbalanced use of chemical fertilisers has adversely affected the soil causing decrease in organic carbon, reduction in microbial flora of soil, increasing acidity and alkalinity and hardening of soil.

b) Excessive use of N-fertiliser are contaminating water bodies thus affecting fish fauna and causing health hazards for human beings and animals.

c) Production of chemical fertilisers adds to the pollution.

To overcome the deficit in nutrient supply and to overcome the adverse effects of chemical cultivation it is suggested that efforts should be made to exploit all the available resources of nutrients under the theme of integrated nutrient management. Under this approach the best available option lies in the complimentary use of Biofertilisers, organic manures in suitable combination of chemical fertilisers. This Integrated approach of nutrient management not only ensures higher productivity but also ensures the good health of the soil and environment. Biofertilisers are essential components of this approach and are being promoted to harvest the naturally available, biological system of nutrient mobilisation.

Benefits

a) Increase crop yield by 20-30%

b) Replace chemical N & P by 25 %
c) Stimulate plant growth
d) Activate soil biologically
e) Restore natural fertility
f) Provide protection against drought and some soil borne diseases.

Types of Biofertilisers

For nitrogen

— Rhizobium for legume crops
— Azotobacter/Azospirillum for non legume crops
— Acetobacter for sugarcane only
— BGA and Azolla for low land paddy

For Phosphorous

— Phosphatika for all crops to be applied with Rhizobium, Azotobacter, Azospirillum and Acetobacter

For enriched compost

— Cellulolytic fungal culture
— Phosphotika and Azotobacter culture

Recommendations

a) For pulses such as moong, urad, arhar, cowpea etc and legume oil seeds such as groundnut and soyabean use Rhizobium + Phosphotika 200 gm each per 10 kg of seed as seed treatment.
b) For non legume crops such as wheat, sorghum maize, cotton mustard etc use Azotobacter + Phosphotika 200 gm each per 10 kg of seed as seed treatment.

c) For Jute - Azospirillum + Phosphotika 200 gm each as seed treatment.

d) Vegetables like tomato, brinjal, chilli, cabbage, cauliflower etc. use Azotobacter/Azospirillum + Phosphotika, 1 kg each as seedling root dip.

e) Low land transplanted paddy Azospirillum + Phosphotika 2 kg each/acre as seeding root dip for 8-10 hrs.

f) Potato, ginger colocassia, turmeric, sugarcane and zoom paddy-use Azospirillum/Azotobacter + Phosphotika 4 kg each/acre mixed with compost and applied as soil treatment.

g) Sugarcane-use Acetobacter + Phosphotika 4 kg each/acre as seed set dipping.

h) Plantation crops-Azotobacter _Phosphotika 4 kg each/acre with compost and applied in soil in two splits per year.

Method of Application

Seed treatment : Suspend 200 gm N biofertiliser and 200 gms Phosphotika in 300-400 ml of water and mix thoroughly. Mix this paste with 10 kg seeds & dry in shade. Sow immediately.

Seedling root dip: For vegetables 1 kg each of two biofertilisers be mixed in sufficient quantity of water. Dip the roots of seedlings in this suspension for 30-40 min before transplanting. For paddy make a bed in the field and fill it with water. Mix biofertilisers in water and dip the roots of seedlings for 8-10 hrs.

Soil treatment: Mix 4 kg each of biofertilisers in 200 kg of compost and leave it overnight. Apply this mixture in the soil at the time of sowing or planting. In plantation crops apply this mixture near root zone and cover with soil.

Precautions

- Store biofertiliser packets in cool and dry place away from direct sunlight and heat.
- Use right combination of biofertilisers
- Rhizobium is crop specific, so use in specified crop
- Do not mix with chemicals
- While purchasing ensure that each packet is provided with necessary information like name of the product, name of the crop for which intended, name and address of the manufacturer, date of manufacture, date of expiry, batch No and instructions for use.
- Use the packet before expiry, only on the specified crop, by the recommended method.
- Biofertilisers are live product and require care in storage. For best results use both nitrogenous and phosphatic biofertilisers. Use of biofertilisers is being empasized along with chemical fertilisers and organic manures. Biofertilisers are not replacement of fertlisers but can supplement their requirement.

BIOFERTILISER UNITS

Sustainable crop production depends much on good soil health. Soil health maintenance warrants optimum combination of organic and inorganic components of the soil. Repeated use of chemical fertilisers destroys soil biota. In nature, there are a number of useful soil micro organisms which can help plants to absorb nutrients. Their utility can be enhanced with human intervention by selecting efficient organisms, culturing them and adding them to soils directly or through seeds. The cultured micro organisms packed in some carrier material for easy application in the field are called bio-fertilisers.

Bio-fertilisers are living microorganisms of bacterial, fungal and algal origin. Their mode of action differs and can be applied alone or in combination. By systematic research, efficient strains are identified to suit to given soil and climatic conditions. Such strains have to be mass multiplied in laboratory and distributed to farmers. They are packed in carrier materials like peat, lignite powder in such a way that they will have sufficient shelf life.

Biofertilisers enhance the nutrient availability to crop plants ; and also impart better health to plants and soil thereby enhancing crop yields in a moderate way. It is a natural method without any problems like salinity and alkalinity, soil erosion etc.. In the vast areas of low input agriculture and oil seeds production, as also in crops like sugarcane, etc, these products will be of much use to give sustainability to production. In view of the priority for the promotion of organic farming and reduction of chemical residues in the environment, special focus has to be given for the production of biofertilisers.

Commercial Prospects

The biofertilisers are mainly purchased by State Agriculture Departments and distributed to the farmers at concessional rates. About 200 to 500 grams of carrier material is only needed per acre, costing about Rs.10/- to 25/-. In view of the above, if the units are selected carefully, there can be assured business. The benefits usually obtained by the use of biofertilisers will not be as visible as that of chemical fertilisers. As the results are not dramatic, many farmers are not aware of the significance, excepting in States like Maharashtra, Gujarat, parts of Karnataka and Tamil Nadu, these are more commonly used with Government's support. In the context of increasing awareness about the use of natural products and organic agriculture, these products will have good scope. Further, the organically grown produces fetch higher prices both in domestic and export markets.

It is estimated that the production of biofertilisers in the country by the existing units is about 7500 to 9000 TPA. This is far below the potential requirement of 7.6 lakh TPA by the year 2000-2001 as estimated by the National Biofertiliser Development Centre (NBDC) Ghaziabad. So far, the Ministry of Agriculture has supported establishment of 67 biofertiliser units in different parts of the country.

Estimated potential Demand for Biofertilisers by 2000-2001

Type of Biofertiliser	*Demand (Tonnes)*
Rhizobium	34,999
Azotobacter	145,953
Azospirillum	74,342
Blue green Algae	251,738
Phosphate solublising microorgaanism	255,340
Total	762,372

This estimated demand of NBDC is based on the cultivated area of the country and treatment of the total seed sown at the rate of 200g biofertiliser per 10 kg of seed. Although this assumption reflects only the macro level requirement, even if 50% of the cultivated area is to be brought under biofertiliser application, there will be a wide gap between the actual production and the requirements. The current trends indicate that there is a steady increase in the demand in the Southern states except Andhra Pradesh, Western States and Madya Pradesh and Rajasthan.

The technology used were indigenous and the scientific aspects of production are standardised by Agricultural Universities and Research Laboratories of GOI. Machinaries and laboratory equipments are available from various manufacturers and are of BIS standards.

Objective of Biofertilser Project

The primary objective of biofertiliser projects could be production of various strains of good quality biofertilisers using most modern technology. The infrastructure and laboratory facilities created, however, can be utilised for the production of bio - pesticides and bio control agents. Multi product range will increase the viability. In line with the technology and objective of biofertiliser production, various facilities are required for the successful implementation of such projects which are indicated below:

Land

It is required to set up laboratory and other facilities and office. Space may also be required for installing tube well / dug well and parking of vehicles. A minimum of ½ acre of land is required for setting up a 150 TPA unit. Preferably, the entire site should be fenced with barbed wire or compound wall with gates at suitable places. The boundary may be planted with thick and tall growing species like Asoka, to filter air and reduce dust.

Layout and buildings

The civil works comprises of factory building for laboratory, Carrier preparation and enrichment, sterilisation, Inoculation and quality control, Maturation of culture, Mixing and packing, storage/ staff etc,. The total covered area of about 3000 sq ft is required for the product manufacturing and other utilities. Rest of the area of land will be enough for future expansion up to 300 to 600 Tonnes per Annum.

Plant and machinery

Manufacture of biofertilisers needs a good number of laboratory equipments as well as other production facilities such as fermentors, culture medium tank, fermentor assembly, autoclaves, boiler, broth dispensers for sterlisation, deminralising plant, air compressor etc,. All the machinery are manufactured in the country. Some of the suppliers undertake the installing the units on a turn key basis.

Manufacturing process and Source of technology

The mother culture of various strains of biofertiliser are supplied from Agricultural Universities and Regional Biofertiliser Development centres (MOA). The unit generally comprises of media preparation room, media store room, inoculation room, growth room, culture transfer room, sterilisation,Mixing and packing, etc. The floor plan should be designed to promote maximum efficiency and minimum contamination. The design should facilitate maintenance of optimum temperature, humidity and ventilation. Inside air of the unit should be free from dust particles.

Infrastructural Facilities for raw material

The raw material required for biofertilser production include ingredients for growth medium for the production of broth, carrier, packing materials like polythene packets. corrugated boxes, etc,.

Utilities

Power

Normally a three phase electric supply is required for these plants. The normal requirements of a 150 TPA unit is about 70 hp. Depending upon the position of power supply, stand by generator may be needed.

Water

A Biofertiliser production unit requires water mainly for steam generation for sterlisation of carrier, broth preparation and cleaning of equipments. Accordingly well/ bore well of designed size and according to the quality of water demineralisation equipments are to be installed. The average per day requirement of water for 150 TPA capacity will be about 2500 to 3000 liters

Compressed air

It will be required for various pneumatic operations as well as for controlled air supply to fermenters, sterlisation / cleaning operations etc,.

Vehicles

The vehicles are required for procurement of carrier material and distribution of biofertilisers as well as for office use. Accordingly one LCV and a jeep have been included in the project.

Manpower

For a unit manufacturing 150 TPA biofertilisers the requirements of manpower is as under :

— 1 Chief Executive Officer
— 1 Chief Biologist / Micro Biologist
— 1 Sales Officer
— 2 Accountant and clerical Assistant
— 2 Drivers
— 1 Floor Supervisor/ Factory Manager
— 2 Technical Staff
— 2-3 Skilled labourers
— 4-5 Semi- skilled, depending upon the volume of production

Unit Size

The size of a biofertiliser unit could be expressed in terms of the capacity of production of various types/ strains of biofertilisers per annum. The projects so far set up in the county vary from 75 TPA to 300 TPA. The size envisaged in the present model is 150 TPA in one shift. The capacity can be easily expanded by adding a few additional equipment like a fermenter and/ or adding another shift.

Environmental Aspects

No hazardous effluents are generated from a biofertiliser unit.

Business Prospects

Considering the importance for organic farming and promotion of sustainable farming practices it is estimated that there will be further scope for adding new units, particularly in the states of Maharashtra, Gujarat, Rajasthan, Madya Pradesh, Tamil Nadu, and Karnataka. The biofertiliser products are picking up mainly in cultivation of sugarcane, pulses, cereals and plantation crops. Since these crops are grown in vast areas, no problem in marketing is foreseen. The units should establish market channels with such niche sectors. They can also have tie up arrangement with government programmes like IPP, NODP etc,.

When a new entrepreneur intends to setup a project, a detailed market survey report is to be prepared. The report may be prepared keeping in view the following aspects. may have to be kept in view:-

1. the present and future demand - Cropping pattern in the supply for various products, gap in supply and expected demand for various products.
2. the major competitors and their present share.
3. the projections of the unit for the next 3-5 years and the basis for projection.
4. the product wise quantities and places where it is to be marketed.
5. the market for the product and type of arrangements for distribution and sales, commission and additional

incentive to be given, the proposed net work and the advertisement plans.

Capital Cost of the Project

Broadly, the capital cost includes the cost of land, development of land, fencing, civil works (plant building, office, godown etc,.) Plant and machinery, preliminary and preoperative expenses etc,. Capital cost of a model biofertiliser unit with a capacity of 150 TPA will be Rs. 73.473 lakhs. The project cost comprises of Rs 2.450 lakhs on land and land development, Rs. 13.500 lakhs on civil structures, Rs. 22.100 lakhs on plant and machinery, Rs. 21.850 lakhs on account of misc. fixed assets, Rs. 3.000 lakhs on preliminary and preoperative expenses, Rs. 7.000 on vehicles, Rs.2.000 lakhs of contingency and Rs 1.573 lakhs of margin money for working capital.

Ministry of Agriculture, Department of Agriculture and Cooperation, Government of India is implementing a central sector scheme viz., "National Project on Development and use of Biofertilisers". Under this project, a subsidy up to Rs.20 lakhs is provided for setting up a biofertiliser production unit of 150 TPA capacity. It is proposed to release the subsidy through back ended subsidy procedure. If the units are getting the subsidy, the amount will be adjusted to the last few instalments of bank loan.

Based on the various techno-economic parameters, the economics of the project have been worked out for the project period or till the repayment of bank loan. The items of income includes sale of biofertilisers. While the expenditure includes the cost of raw material, transportation and commission, power, fuel packing distribution, wages and salary, repairs and maintenance, insurance, advertisement and other overheads. The income as well as expenditure for each year are worked out and subiected to cash flow analysis.

The cash flow statement covering the Benefit Cost Ratio (BCR), Net Present Worth (NPW) and Internal/financial rate of return (IRR/FRR) have been worked out for the project. Normally the BCR should be greater than 1, NPW should be positive and IRR/FRR should be greater than 15%. For the model project under consideration, the BCR is 1.132, NPW is Rs. 22.276 lakhs and IRR is more than 15%.

The projects on manufacturing biofertiliser products would be considered for refinance support by National Bank. Therefore, all participating banks may consider financing this activity subject to their technical feasibility, financial viability and bankability.

Lending Terms

— *Margin money:* The promoters/company should normally meet 25% of the project cost out of their own resources.

— *Interest rate:* Interest rate will be determined by RBI/ NABARD from time to time.

— *Security:* As stipulated by the RBI.

— *Repayment period:* Depends upon the gross surplus generated, it may be up to 9 years with first one/two years as the grace period.

— *Refinance assistance:* NABARD provides refinance assistance as per the existing pattern from time to time.

— *Subsidy:* A subsidy ofd 25 % of the project cost will be given as subsidy by Ministry of Agriculture.

TECHNICAL ASPECTS OF BIO-FERTILISERS

Bio-fertilisers, in strict sense, are not fertilisers which directly give nutrition to crop plants. These are cultures of micro organisms like bacteria, fungi, packed in a carrier material. Thus, the critical input in Biofertilisers is the micro organisms. They help the plants indirectly through better

Nitrogen (N) fixation or improving the nutrient availability in the soil.

The mode of action depends on the species of the organism. Some agents like Rhizobium cultures enhance N fixation in legumes by imparting effective modulation as they are symbiotic bacteria living in association with leguminous plants.

There are free living bacteria like Azotobacter when applied to soil, enhance the N availability. There are certain other organisms which act on the soil minerals and dissolve the native nutrients like P which is otherwise not readily soluble. The most commonly produced and marketed biofertilisers are

1. Rhizobium - 3 to 4 strains
2. Azotobacter
3. Azospirillum
4. Phosphate Solubulising Bacteria - 2 species.
5. Blue Green Algae and Azolla - on-farm level.
6. VAM - in a limited way.

The critical factors which are responsible for the effectiveness of a particular bio-fertiliser are as follows:

— Suitability of the species to the target crop
— *Suitability of the strain* : There are specific strains of rhyzobium for different leguminous species like Cowpea, Redgram, Soybean, Alfalfa etc. Biofertiliser of specific culture should be used for specific crop.
— Identification of strains as suited to the agro-eco system, particularly the soil P^H and moisture conditions. Through research, specific strains as suited to a particular soil and environmental conditions are usually identified and pure mother cultures are maintained in research labs for supply to the commercial manufacturers, e.g. germ plasm of Rhisobium cultures is maintained at IARI, New Delhi.

— The aseptic conditions of manufacturing, the cell count of living organism present in the carrier material, purity and level of contamination.

— The conditions of carrier material in which the culture is packed and the quality of the packing material, which determine the shelf life.

— The conditions in which the packed materials are stored, distributed and kept with the farmers before it is applied.

— Soil conditions particularly P^H, organic matter content and moisture level; and agronomic practices.

The benefits usually obtained will not be as visible as that of chemical fertilisers except in some critical conditions. Biofertilisers can add nitrogen from 20 kg/ha to 200 kg/ha depending upon the optimum conditions. Pastures and forages respond more than grain crops. The yield increases usually range around 10-35%. However, in the vast areas of low input agriculture and in the context of imparting sustainability to crop production at reduced chemical pollution, this product will be of much use. Rhizobium culture treatment becomes essential when new types of legumes like Soybeans are introduced in new areas.

Application of biofertilisers should not be viewed from the only angle of nutrient supply to the crops. They add life to the soil rendered sterile by the excess use of chemicals, etc. Some of them possess growth promoting substances and also reduce the incidence of certain diseases. These inputs are crucial if some one would like to take up organic farming.

REFERENCES

Altieri, Miguel A., *Agroecology: The science of sustainable agriculture,* Westview Press, Boulder, CO.

Dore, J., *Sustainability Indicators for Agriculture: Introductory Guide to Regional/National and On-farm Indicators,* Rural Industries Research and Development Corporation, Australia, 1997.

Gold, Mary, *Sustainable Agriculture: Definitions and Terms*, Special Reference Briefs Series no, SRB 99-02 Updates SRB 94-05 September 1999, National Agricultural Library, Agricultural Research Service, U.S. Department of Agriculture, 1999.

Jahn, GC, B. Khiev, C. Pol, N. Chhorn, S. Pheng, and V. Preap, "Developing sustainable pest management for rice in Cambodia", In S. Suthipradit, C. Kuntha, S. Lorlowhakarn, and J. Rakngan [eds.] "Sustainable Agriculture: Possibility and Direction" Proceedings of the 2nd Asia-Pacific Conference on Sustainable Agriculture 18-20 October 1999, Phitsanulok, Thailand. Bangkok (Thailand): National Science and Technology Development Agency, 386 p., 2001.

Lindsay Falvey, *Sustainability - Elusive or Illusion: Wise Environmental Management*, Institute for International Development, Adelaide pp259, 2004.

Hecht, Susanna and Alexander Cockburn, *The Fate of the Forest: developers,* destroyers and defenders of the Amazon. New York: Verso, 1989.

Netting, Robert McC., *Smallholders, Householders: Farm Families and the Ecology of Intensive, Sustainable Agriculture*, Stanford Univ. Press, Palo Alto, 1993.

7

Mycorrhizal Biofertilisers

A mycorrhiza is a type of endophitic, biotrophic, mutualistic symbiosis prevalent in many cultivated and natural ecosystems. There are three major groups of mycorrhiza: Ectomycorrhiza, Ectendomycorrhiza and Endomycorrhiza. Ectomycorrhiza and endomycorrhiza are important in agriculture and forestry. In Thailand, endomycorrhiza biofertiliser has been investigated for ten years. Initially the mycorrhizal biofertilizer production is for economic crops such as fruit trees. Now the biofertiliser can be used for vegetables and rubber.

Endomycorrhiza (vesiclular arbuscular mycorrhiza; VA mycorrhiza; now known as arbuscular mycorrhiza, AM) play a very important role on enhancing the plant growth and yield due to an increase supply of phosphorus to the host plant. Mycorrhizal plants can absorb and accumulate several times more phosphate from the soil or solution than non–mycorrhizal plants. Plants inoculated with endomycorrhiza have been shown to be more resistant to some root diseases.

Arbuscular Mycorrhizal (AM) fungi (or Vesicular-Arbuscular Mycorrhizal, VAM fungi), belonging to the Phylum Glomeromycota are symbionts with terrestrial plant roots. It is now generally recognised that they improve not only the phosphorus nutrition of the host plant but also its growth, which may result in an increase in resistance to drought stress and some diseases. Therefore, AM fungi offer a great potential for sustainable agriculture, and the application of AM fungi to agriculture has been developed. In fact, in some countries the AM fungal inocula have been commercialised. Since it is laborious and cost-consuming for production of AM fungal inocula because of their obligate biotrophic nature, the ways to increase the function of the indigenous AM fungi in soil have also been developed.

BENEFITS

Mycorrhiza plays a very important role on enhancing the plant growth and yield due to an increase supply of phosphorus to the host plant. Mycorrhizal plants can absorb and accumulate several times more phosphate from the soil or solution than non–mycorrhizal plants. Plants inoculated with endomycorrhiza have been shown to be more resistant to some root diseases. Mycorrhiza increase root surface area for water and nutrients uptake.

The use of mycorrhizal biofertiliser helps to improve higher branching of plant roots, and the mycorrhizal hyphae grow from the root to soil enabling the plant roots to contact with wider area of soil surface, hence, increasing the absorbing area for water and nutrients absorption of the plant root system. Therefore, plants with mycorrhizal association will have higher efficiency for nutrients absorption, such as nitrogen, phosphorus, potassium, calcium, magnesium, zinc, and copper; and also increase plant resistance to drought. Benefits of mycorrhizal biofertilise can be seemed as follows:

1. Allow plants to take up nutrients in unavailable forms or nutrients that are fixed to the soil. Some plant nutrients, especially phosphorus, are elements that dissolve were in water in neutral soil. In the extreme acidic or basic soil, phosphorus is usually bound to

iron, aluminum, calcium, or magnesium, leading to water insolubility, which is not useful for plants. Mycorrhiza plays an important role in phosphorus absorption for plant via cell wall of mycorrhiza to the cell wall of plant root. In addition, mycorrhiza help to absorb other organic substances that are not fully soluble for plants to use, and also help to absorb and dissolve other nutrients for plants by storage in the root it is associated with.

2. Enhance plant growth, improve crop yield, and increase income for the farmers. Arising from improved water and essential nutrients absorption for plant growth by mycorrhiza, it leads to improvement in plant photosynthesis, nutrients translocation, and plant metabolism processes. Therefore, the plant has better growth and yield, reduce the use of chemical fertiliser, sometimes up to half of the suggested amount, which in turn increases income for the farmers. As in the trial involving mycorrhizal biofertiliser on asparagus it was observed that, when the farmers used suggested amount of chemical fertiliser together with mycorrhizal biofertiliser, it was found that the crop yield improved by more than 50%, and the farmers' income increased 61% higher than when chemical fertiliser alone was used.
3. Improve plant resistance to root rot and collar rot diseases. Mycorrhizal association in plant roots will help plant to resist root rot and collar rot diseases caused by other fungi.
4. It can be used together with other agricultural chemicals. Mycorrhiza are endurable to several chemical substances; for example; pesticide such as endrin, chlordane, methyl parathion, methomyl carbofuran; herbicide such as glyphosate, fuazifopbutyl; chemical agents for plant disease

elimination such as captan, benomyl, maneb triforine, mancozed and zineb.

Arbuscular Mycorrhizal Fungi

Taxonomy of AM fungi

AM fungi show the peculiar characteristics in morphology and physiology. Spores of AM fungi are generally formed in soil and their sizes (50-500 μm in diameter) are much larger than those of other fungi. There is no septum in their hyphae. No sexual growth-phase has been observed. Spores germinate when they are under favourable conditions, extend their hyphae and colonised plant roots.

The fungi penetrate the hyphae into cortex layer of roots and form the hyphal organs, "vesicles" and "arbuscules" which are characteristics to AM fungi. AM fungi belonging to Gigasporaceae are known not to form vesicles. Colonisation on plant roots is essential for proliferation of AM fungi. AM fungi are thus recognised as obligate symbiotic fungi.

The interaction between AM fungi and plants is generally mutualism based upon nutrient exchange. Because of morphological characteristics such as no hyphal septum, AM fungi had long been recognised as a member of Zygomycota. Recent molecular phylogenetic studies showed that Zygomycota is poly-phyletic and that AM fungi should be separated from other Zygomycota.

A new Phylum Glomeromycota has been proposed for AM fungi. Current classification system is summarized in Figure 1. This classification is mainly based upon the sequence data of rRNA gene. However, some new genera have been raised with relatively small numbers of isolates, so further study may revise the present classification system.

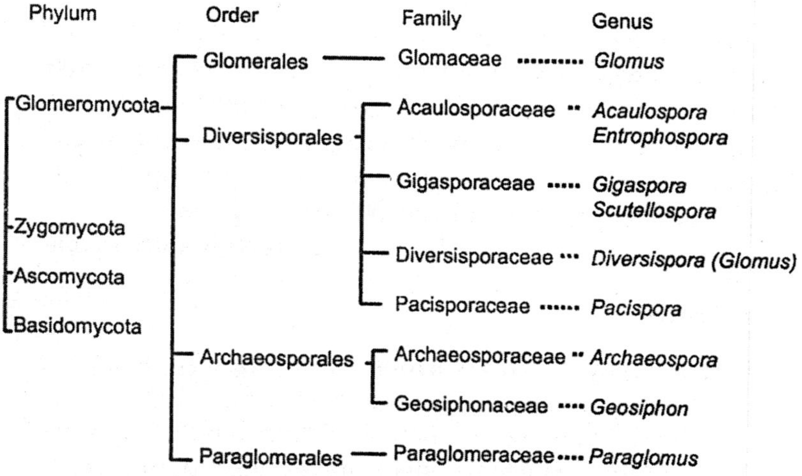

Figure 1: Taxonomy of arbuscular mycorrhizal fungi.

Arbuscular mycorrhizal fungi in roots

Arbuscular mycorrhizal fungal structure in roots is usually not observed without appropriate staining. Freshly collected root samples should be washed gently and be free from soil particles. Ultrasonic treatment is effective to disperse soil particles closely adhered to roots. Roots are treated with 10 % KOH solution for 30 min to 1-2 hours in a hot bath, depending on thickness of root structure.

Treated roots are washed with water and treated with 2 % HCl solution. Acidified root samples are stained with 0.05 % trypan blue in lactic acid for 10-15 min in a hot bath or for a few hours without heating. The roots are destained with lactic acid or lacto-glycerol and are now ready for microscopic observation. The stained roots may be observed first under a dissecting microscope with transmitted illumination and then observed under a compound microscope. Fungal structures are stained and can be easily recognised.

Isolation of spores from soils

Spores of AM fungi in soil can be collected by the wet sieving method. The gravity of spores is a little lighter than that of soil particles. Successive decantation of soil suspension followed by sieving with fine mesh can concentrate the spores from soil. Since the spores are globular or sub–globular in 50–500 μm in diameter, they, in sievings can be recognised under a dissecting microscope

Equipments

1) *Sieve:* Sieves with various mesh size. At least the following mesh size are required; 1 mm, 100 μm and 50 μm. Other sizes such as 500μm and 250μm are preferable. Stainless steel sieves are commercially available. However, it is possible to make a plastic sieve with PVP tubes and nylon mesh by yourself.
2) *Fine glass pipettes:* Tip of disposable glass Pasteur pipette (1 ml) is softened with flame of gas burner and is sharpened. Various sizes of tips fitting to sizes of spores can be prepared.
3) *Forceps:* Light fine tweezers is preferable. Tweezers Model 113SA (OHM-Werkzeuge, Germany) is the most convenient with good handling.
4) *Dissecting microscope*: Stereoscopic zoom microscope with bifurcated illuminator of fiber arm is advisable. Transmitted illumination system is also needed.
5) *Compound microscope*: Biological compound microscope is needed. Nomarsky's DIC illumination system is advisable.

Procedure

Ten to 50 g of freshly collected soil sample is put into 1 to

2 litters of plastic beakers. Usually rhizosphere soils are rich in AM fungal spores. Beaker size can be changed depending on the soil sample size. Soil is suspended with about 500 ml to 1 litter of tap water. Soil macro-aggregates should be crushed with hand. After 10-30 seconds of settling down of soil particles, the upper layer of soil suspension is poured into the sieving (Figure 2). The procedure should be repeated until the upper layer of soil suspension is transparent.

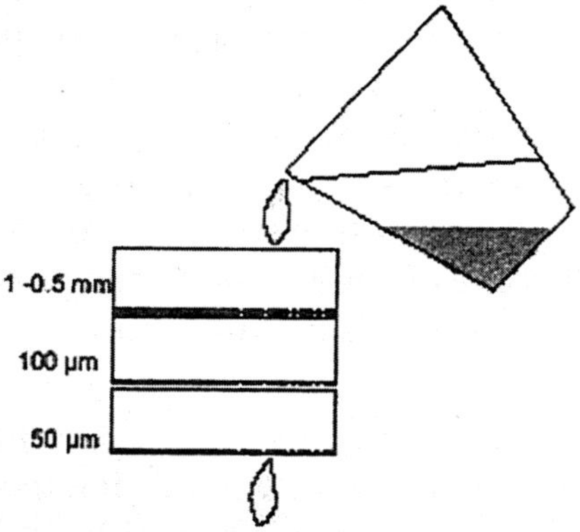

Figure 2: Sieving of soil suspension.

The sievings on the fine mesh is collected into a small beaker and dispersed with ultra sonication. Weak sonication is enough, and strong sonication may destroy fungal spores. Then the dispersed sample is again passed through the sieve. Depending on toughness of soil aggregate, the sonication process can be repeated. Usually AM fungal spores are collected on 100 mm. Some small spores are on 50 mm. To collect large spores such as *Gigaspora margarita,* 250 mm sieve is efficient.

Spores of AM fungi have characteristic shapes and colours, and so it is not difficult to discriminate the spores in organic debris collected on the sieves. However, it is recommended that those who have not yet observed the spores before should learn from the experts how the spores look like. In the soil such as grassland soil rich in organic debris, it may be hard to find the spores hidden by the debris. In such a case, sucrose density centrifugation technique is often used to separate spores from the organic debris.

Morphological observation of spores

Morphology of spores is a basis for identification of AM fungi, because the hyphae and the organs such as arbuscules and vesicles are not specific to species. Spores collected from soil often deteriorate so that they may be used only for tentative identification at genus level. For detailed observation, culturing the target AM fungus is required, and the spores from pot culture should be used. At least 30-50 spores from the same morphological spore type should be observed, and more observation is recommended.

1) *Observation of intact spores under dissecting microscope:* Spores collected from soil or culturing medium are put in a watchglass or a small Petri dish, and their shape, colour and the attachment to spores are observed. Spores should be classified into each spore type based upon morphology. For each spore type, detailed observation is conducted. For colour description, standard colour chart such as soil colour chart or "colour chart of glomalean fungi" should be used. The colour chart should be under the same illumination as used for spore observation, because the colour itself is greatly affected by the characteristics of illumination. It should be reminded that hyphal attachments such as sporiferous saccule for Acaulosporaceae and subtenting hyphae for Gigasporaceae are often lost during collection of spores from soil.
2) *Observation of spores mounted on slide glass under a compound microscope:* Spores are mounted with

polyvinyl lactoglycerol (PVLG) on a slide glass. Several slides should be made. These are for intact spores mounted with PVLG, for crushed spores mounted with PVLG, for spores mounted with PVLG containing Meltzer's reagent.

The morphological characteristic as indicated in Table 1 should be recorded. These characteristics may be helpful to identify genus of the target fungus (Figure 3). However, some genera such as *Archaeospora* needs not only these morphological characteristics but also sequence data. For species identification, the characteristics should be compared with those in the species description in the original reference. Species identification without enough expertise may cause trouble in scientific society, so those who wish to identify species of AM fungi should consult the experts in AM fungal taxonomy.

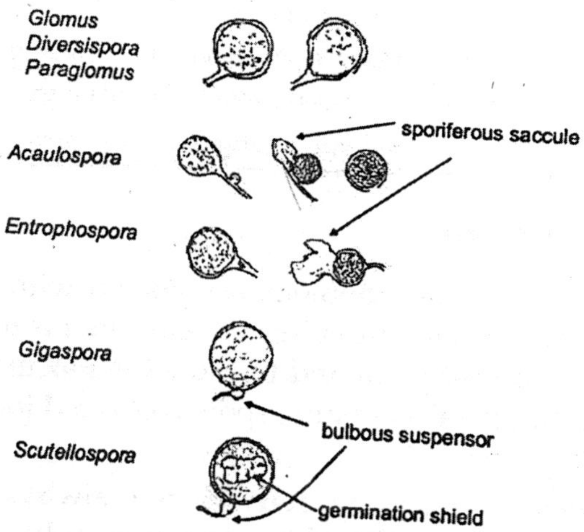

Figure 3: Morphology of representative genera of arbuscular mycorrhizal fungi.

For spore size, at least 40-50 spores should be examined, and more examination is preferable. PVLG (poly-vinyl lacto-glycerol):

Polyvinyl alcohol (polymerisation 1000-1500), 1.66 g, is dissolved in 10ml of deionised water. Complete dissolution may need 6 hours at 80 °C. The dissolved polyvinyl alcohol is mixed with 10 ml of lactic acid and 1ml of glycerol. It can be used more than a day after preparation.

Table 1. Morphological character of spores of AM fungi

Shape:	(i.e. globular, spherical, irregular etc)
Size:	Globular: diameter (minimum - average maximum)
	Irregular shape: length x width (minimum average – maximum)
Colour:	(compare with Standard Colour Chart)
Hyphal attachment:	(i.e. sporiferrous saccule, bulbous suspensor etc)
	sporiferrous saccule = Acaulospora, Entrophospora, Archaeospora bulbous suspensor = Gigaspora, Scutellospora
Auxiliary cell:	(presence = Gigaspora, Scutellospora, none)
Sporocarp:	(presence, none)
Germination shield:	(presence =Scutellospora, absence)
Surface ornamentation:	(i.e. smooth, rough, reticulate etc)
Vesicle:	(presence or absence in mycorrhizal roots)

Culturing AM fungi

AM fungi need the symbiotic association with plants for proliferation. Therefore, culturing AM fungi is to inoculate AM fungi to host plant and to grow the inoculated plant. For the AM fungal inoculum, spores collected form soil can be used.

However, spores in soil are not always active in colonising plants. Therefore, trapping culture is often employed. Soil or sieving of soil is used as inoculum (Soil Trap Culture). To isolate AM fungi colonising roots, mycorrhizal plants collected from field can be transplanted to potting medium as Plant Trap Cultur (Figure 4).

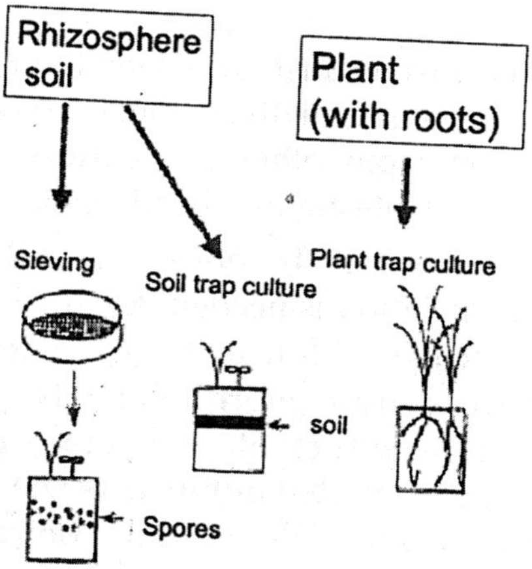

Figure 4: Methods for trapping AM fungi.

— *Potting medium:* Sterile soil or soil-sand mixture is usually used. We prefer to use commercially available "Akadama-tsuchi" which is collected from subsoil of volcanic ash soil and is prepared for horticulture use. Various potting materials for horticulture can be also used. However, the materials for potting medium should be low in available phosphate and preferably not rich in organic matter. In some cases the fungi isolated from some specific soils may need the specific soil properties for their growth.

— *Host plant:* Various mycotrophic plants can be used: leguminous species and grass species, and other herbaceous species. Onion and leek are also good hosts. AM fungi generally do not show host specificity but some species show host preference. Therefore, the plant species from which the target AM fungus is isolated can be used as a host plant.

— *Growth conditions:* Any conditions, which support good growth of host plants, are acceptable. To avoid contamination, a growth chamber is preferable. If greenhousē is used, it should be kept clean. It should be reminded that cross-contamination or contamination from dust is inevitable under open-air conditions, even in growth chamber. To prevent cross-contamination from other pot culture in the same chamber, use of plastic bag is advisable.

— *Single spore isolation:* To purify an isolated fungus, single spore isolation is needed. Even if the spores are morphologically identical, it often contains contaminants whose morphology is very similar. Successive pot culture of such multispore isolates would cause unexpected outbreak of the contaminant. Furthermore, even if the culture contains only one species, it may be composed of genetically diverse populations. For such a genetic studies or population genetics, the purification through single spore isolation is essential.

For single spore isolation, no specific equipment is needed. For efficient handing, two sets of dissecting microscopes are place side by side. One microscope is for picking up single spore from spores in a dish. Another is for inoculation of a spore on roots. Seedling placed in the pot is placed under another microscope. Under the first microscope, single spore is picked up and transferred under the second microscope.

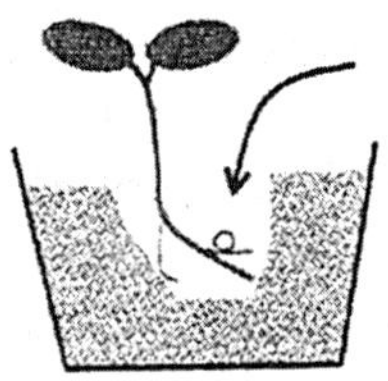

Figure 5: Single spore isolation.

Under the second microscope, the spore is placed on fine roots or root tip of the seedling (Figure 5). If culture is successful, the detailed morphological observation is required. Potting medium can be dried by stopping watering to the pot. After the host plant wilt, the dried soil containing spores can be stored for a year at 4-5°C. It is advisable that the isolated fungi are re-cultured every year.

DNA extraction from spores

Molecular phylogenetic information is essential for taxonomy of AM fungi. Sequence data for conserved genes such as rRNA is obtained by PCR amplification followed by sequencing with DNA extracted from spores. Many protocols for DNA extraction are reported.

Cleaning of spores

Clean spores are collected with tweezers or fine Pasture pipette. Spores should wash several times with sterile water with sonication for 10-20 seconds. If spores show water repellency, surfactant such as Tween 80 can be used. Microplate with 6 or 12 wells is convenient for successive washing.

Crushing spores

We use a cap of Eppendorf tube. The cap is cut from the tube and placed upside down on the stage of a dissecting microscope. Twenty µl of Instagene (Biorad) is added to the cap. A spore or spores are put into the cap and crushed thoroughly with a micro-pestle or any fine rod. The tube itself is put on the cap with crushed spores. Then the tube is centrifuged for while for spinning down the reagent with crushed spores from the cap into the tube.

Extraction of DNA

Follow the instruction provided by manufacture. Further

purification of DNA with ethanol/chloroform precipitation is sometimes needed to remove inhibitor for PCR amplification.

PCR amplification and DNA sequencing

Conditions of PCR amplification depend on the primers you will use. AM fungi are multi-nuclear organisms and often show polymorphisms in their sequence. It is advisable to sequence several clones from the target fungus and to check the phylogenetic position of the sequence by constructing phylogentic tree if the target genes are located within a reasonable clade of the tree.

REFERENCES

Brundrett, M. *et al.*, *Working with Mycorrhizas in Forestry and Agriculture*, Australian Centre for International Agricultural Research, 1996.

Norris, J.R., Read, D.J. and Varma, A.K. (ed.)., *Methods in Microbiology, Vol. 24, Techniques for the Study of Mycorrhiza*, Academic Press, 1992.

Sieverding, E., *Vesicular-Arbuscular Mycorrhizal Management in Tropical Ecosystems*, Deutche Gesllshaft für TEchnische Zusammnearbeit (GTZ) GmbH, 1991.

8

Utilisation of *Azolla* as a Biofertiliser

In order to have sustainable agriculture, a constant supply of nutrients with high crop productivity is needed. For farmers to increase productivity, they must apply methods that replenish soil nutrients removed by crops. This should be on a long term basis so as to sustain farming systems. Among the factors affecting productivity, nutrient availability is of prime importance because of its effect on key physiological and developmental processes that determine plant growth.

Fertilisers are utilised to improve the capacity of the soil to supply the necessary nutrients in an agricultural system and are therefore of great importance. In the less developed countries, many farmers cannot afford inorganic fertilisers. This has led to interest in biofertilisation with emphasis on biological nitrogen fixation. For centuries, the red water fern *Azolla filiculoides* Lam. (Azollaceae) has been utilised as a green manure for lowland rice in Vietnam and China. Due to a symbiotic relationship with a cyanobacterium, *A. filiculoides* is able to fix atmospheric nitrogen (N) which makes it a potential biological source of N and therefore economically important.

Use of this small fern would be appropriate, cheaper, and increase productivity in agriculture. World wide interest in its use gained momentum in the 1970's as a result of the fuel crisis that threatened the production of inorganic

N fertiliser. Research carried out on different species and strains of *Azolla* showed increased yields in rice production under flooded conditions. In South Africa, *A. filiculoides* occurs in fresh water bodies from where it can be harvested and utilised by farmers. This provides a basis to investigate possible use of *A. filiculoides* under dryland conditions in South Africa.

AZOLLA FILICULOIDES

Many biological studies on the identification, agronomic use and control of *A. filiculoides* have been carried out. In this study, the key point is the beneficial effects of *A. filiculoides* as an organic fertiliser. Thus information on where it occurs and how it provides N is necessary so that harvests are made at optimum times for maximum nutrient input into the soil.

A. filiculoides is a small free floating fresh water heterosporous fern. It belongs to the order Salviniales and is divided into two sections. Section Rhizosperma with three floats includes *A. pinnata* R. Br. and *A. nilotica* Decne. ex Mett. while section Euazolla with nine floats includes *A. filiculoides* Lam., *A. mexicana* Presl., *A. caroliniana* Willd. and *A. microphylla* Kaulf. *A. filiculoides* has a branched rhizome with alternate leaves and roots that hang into the water. Each leaf consists of two lobes; an aerial dorsal lobe (chlorophyllous) and a partially submerged ventral lobe (colourless and cup shaped for buoyancy).

Symbiosis

It is of interest to note that *A. filiculoides* contains cyanobacteria that make it rival with legumes in N production. A symbiotic relationship exists between an endophytic cyanobacterium (*Anabaena azollae* Strasburger) and *A. filiculoides*. The cyanobacterium lives within the dorsal leaf cavity of the fern. The endosymbiont fixes N for both organisms while the fern provides a protected environment and a fixed carbon source that benefits the cyanobacteria.

Anabaena azollae excretes ammonium ions into the leaf cavity. These are assimilated by the fern using glutamine synthetase and converted into amino acids. The *Azolla-Anabaena* relationship was found to fix N better than *Sesbania rostrata* (a legume) and able to release it for rice uptake. This was at a minimum of 30-40 kg N ha^{-1} to 70-110 kg N ha^{-1}. According to Talley and Rains, the highest N yield by the fern is before sporulation and therefore plants should be harvested then for optimum use in agriculture.

Reproduction

In view of the fact that *A. filiculoides* is a weed, any use for it has to be balanced against the possibility of further spreading of the weed. It is important to note that *A. filiculoides* reproduces both sexually and asexually and its symbiotic relationship exists throughout all generations. At the first leaf initial of the ventral lobe, the fern produces two sporocarps. Each megaspore contains a colony of *Anabaena azollae* in the form of akinetes which colonise the sporophyte.

Fertilisation occurs within the water and the embryo develops into a sporophyte that floats to the water surface after the growth of the first and second leaf. Sporulation is associated with dense mat formation in that spores are only formed when the fronds are multi-layerered. Asexual reproduction is by fragmentation of the fronds. The mature secondary rhizomes or branches form abscission layers at their bases and these secondary rhizomes break off from the main rhizome. Under suitable field conditions, it can double its mass every 3-5 days.

Growth

A. filiculoides is capable of photosynthesising at rates higher than most C4 plants due to a variety of light harvesting pigments in the two symbionts. The *Anabaena azollae* contains cells known as heterocysts that are specifically responsible for N fixation. It has been found that in non symbiotic species of *Anabaena*, the heterocyst frequency never makes up more than 3 to 5% of the cells and yet in mature *Azolla* leaves, they compose up to more than 30% of the cell numbers of the symbiont. Various environmental factors affect the growth of the *Azolla* plant; irradiance, temperature, water depth, pH, nutrient status and heavy metals. It grows exponentially at a temperature range of 10°C up to 25°C.

The plant may turn from green to red which is thought to be a result of production of anthocyanins due to direct sunlight, low temperatures as shown in *A. filiculoides* and a lack of nutrients especially P. Strong sunlight has been shown to impair the light harvesting mechanism in *A. caroliniana* resulting in slow N fixation. All these studies show that, like all other plants, the growth of *Azolla* is reduced under poor environmental conditions which in turn reduces the amount of N available in the fern, making it less efficient as a biofertiliser.

Source of a Biofertiliser

It is advantageous to have a readily available source of a biofertiliser which saves on costs of labour in planting. *A. filiculoides* is widely distributed as a weed in South Africa especially on still and slow moving fresh water bodies. Originally it was found from southern South America through western North America to Alaska. *Azolla* thrives in tropical, subtropical and warm temperate regions. It seems to have been introduced to S. Africa as an ornamental during the nineteenth century.

In South Africa it was first recorded in the Oorlogspoort River near Colesburg in 1948 and has since been recorded at 64 localities by the National Botanical Institute. In South

Africa, *A. filiculoides*, produces dense mats (5-30 cm thick) which result in reduction of drinking water quality due to bad odours, coloration and turbidity. It increases water borne, water based and water related diseases. It increases evapotranspiration and reduces the water surface area available for recreation and transport.

The aquatic biodiversity is also reduced with a specific example being the danger to the Eastern Cape Rocky, *Sandelia bainsii* in the Eastern Cape. Beside these factors, it also clogs irrigation pumps, drowns livestock and reduces water flow in irrigation canals. Therefore any available means of reducing this environmental hazard is important.

A method that is able to reduce the weed as well as utilise it would be ideal. Although *A. filiculoiides* is a weed, the fact that it is a source of N, K, Ca and other nutrients such as Iron and Copper makes an investigation of its use for upland (dryland) agriculture worthwhile. It should also be considered as an alternative means of improving soil fertility with longterm effects, in view of the threat of inorganic fertilisers to the environment.

Controlling *A. filiculoides*

One of the possible means of controlling *Azolla* especially on small water bodies such as dams is mechanical harvesting. Not only is this a control mechanism but it also makes the harvested fern available for biofertilisation. Other possible means of control are biological and chemical. Hill states that mechanical control is ecologically sound whereby infestations of the weed in small accessible areas can be removed with rakes and finely meshed nets. However this facilitates reinfestation under ideal conditions and is labour intensive.

Biological control is a sustainable and ecologically benign method. To ensure effective biocontrol, a host-specific organism is needed. In late 1995, the weevil *Stenopelmus rufinasus* was imported from Florida into South

Africa and tested for host specificity by the Weeds Division of Plant Protection Research Institute, Pretoria. In its country of origin, the weevil has not been recorded on any other plant species apart from *A. filiculoides*. Research was carried out on 27 plants in 16 families. Only *A. filiculoides* was found capable of supporting populations of this weevil.

The Directorate of Plant and Quality control of the Department of Agriculture was permitted to release it as a natural enemy for the weed. In December 1997, the weevil was released in South Africa as a biological control for *A. filiculoides* on a dam in the Austin Roberts Bird Sanctuary in Pretoria. Within two months, the weevils had caused the *Azolla* mat to collapse and after one year there has been no resurgence of the plant even from spores. Over the period of January 1998 and March 1999, weevil release at 46 sites all over South Africa, resulted in control of *Azolla* mats at 20 sites and reduction in infestation at five sites.

Biofertilisation

Chemical fertilisers

An applicable method of increasing yields is the use of chemical fertilisers which are expensive, disturb the equilibrium of agroecosystems and pollute the environment. This includes pollution of water systems and ground water due to runoff which contains nitrates, acidifies soils and reduces microbial activity. Most of the chemical nitrogenous fertiliser is produced by industrial N fixation. Each unit of N fertiliser produced requires two units of petroleum. This is expensive, especially for the farmers in the less developed countries, and petroleum is a non-renewable resource.

Organic fertilisers

In comparison, the application of biofertilisers is

inexpensive as it makes use of freely available solar energy, atmospheric N and water. Biofertilisers may be microbial inoculants or microbially-converted organic materials which are used to supply nutrients to plants. It is well known that nutrient availability in soil contributes to plant fitness in terms of shoot and root productivity. Concern for the protection of our environment has led to increasing interest in the use of biofertilisers. They are renewable resources, non pollutants, supply other nutrients and improve the general fertility of the soil.

Biofertilisers promote self sufficiency and reduced costs for resource-poor farmers. One of the most limiting elements in agriculture is N and its addition to soil contributes to the dry mass of plants in the form of proteins. Most cultivated soils are deficient in N and yet the three most important cereals need 20-40 kg soil N ha^{-1} for a period of 3-5 months. Agren, on the assumption that plant growth is determined by the amount of N in the plants, developed a theory on the concept of N productivity in relation to plant nutrition which states that N exerts a strict control over growth and is therefore an essential nutrient.

Mineral nutrients

Apart from N other mineral nutrients are required for plant growth and an increase of these from the deficiency range will increase the growth rate and yield, although the response increases to an optimum after which it reverses and mineral availability interacts with water availability.

The number of seeds or fruits and flower initiation can be affected by mineral nutrition which is clearly the case with various micronutrients. In wheat, copper deficiency affects the reproductive phase as it inhibits anther formation, although it hardly affects the straw (culms and leaves). Boron is essential for pollen tube formation and low supply inhibits flowering and seed production.

A limited supply of mineral nutrients such as N, K, P and Mg that are available for re-translocation from source to sink may affect grain yield rather than limited carbohydrate source. Concentrations of Zn and Fe in cereal grain increase with increase of fertiliser. N, P, and K increases shoot and root development and uptake of other nutrients.

Azolla may supply micronutrients such as Zn to plants if used as a slow release organic fertiliser. Short term experiments conducted in India, in the wet season, when inorganic fertiliser and green manure *Azolla* were combined, showed that the rice yield and agronomic efficiency was greater than for inorganic fertiliser alone.

Utilisation of *Azolla*

Azolla been utilised as a green manure in rice paddies in China and Vietnam for over 2000 years. It is reported to have been first domesticated and used by a peasant woman (Balteng) in Lan Van village in Vietnam. During the mid 1960s, the Vietnamese government renewed efforts to extend the area under *Azolla* cultivation. In China it was reported that *Azolla* as a green manure decreased specific gravity and increased porosity and organic matter in lowland soils. *Azolla* is applicable as a biofertiliser because of its endogenous supply of fixed N.

The *Azolla-Anabaena* symbiosis has a high productivity and fixes N at high rates which makes it applicable as a biofertiliser in agriculture to supply a natural source of crucial N. The symbiont is able to release nutrients into the soil availing them for plant uptake. As fresh material, it has mainly been applied to lowland rice and it has been found to improve soil structure, decompose rapidly, accumulate K in low K environments and grow rapidly. *A. filiculoides* when decaying releases nutrients better as fresh matter than desiccated matter in water. The decomposition rate depends on the C/N ratio, temperature and soil properties.

The *Azolla* biomass is readily decomposed in flooded rice conditions. It has a favourable C:N ratio of 8-17. The addition of organic matter to the soil narrows the C/N ratio. Ram *et al.* showed that addition of *Azolla* decreased pH, increased physical soil properties such us aggregation of soil particles, soil structure and permeability leading to better water holding capacity and less evapotranspiration in sandy loam rice soils.

When compared to inorganic fertilisers, the green manure had positive residual effects. The application of *Azolla* increases the N, P and K content of the soil. In Niger the addition of fresh *A. pinnata* to soil increased the tiller, panicle and grain yield of rice. In a 2 year experiment, carried out in India, *Azolla* (40 kg N ha^{-1}) mixed with *Sesbania* (40 kg N ha^{-1}) was applied to a low land rice-wheat cropping with increase in yields equivalent to the application of 80 kg urea N ha^{-1}.

Lumpkin and Plucknett state that the introduction of *A. filiculoides* to China greatly improved rice production with the addition of 30-40 N ha^{-1}. This would be ideal for situations where low cost is required, such as saving on the use of commercial fertilisers. It is important to note that the use of biological N fixation will be affected by environmental factors and the effective use of available N by the crops. Robinson *et al.* found that the nutrient uptake by a crop depends on the extent to which the supply of ions from the soil can match the demand for nutrients created by growth.

The gradual release of nutrients by decomposition of organic fertilisers is suitable for crop growing and does not have the leaching problems associated with inorganic fertilisers used on sandy low organic matter soils. Lumpkin and Plucknet state that the addition of *Azolla* benefits poor soils more than good soils. Talley and Rains state that dried *Azolla* increased yields to the same degree as application of ammonium sulphate at the rates of 40 and 80 kg N ha^{-1}.

In a 3 year study, Kolhe and Mittra showed that *Azolla* can be utilised as a substitute to inorganic N with a residual effect of 63% on wheat over the control in a rice-wheat cropping. Ram and Prasad found that application and incorporation of *Azolla* (60, 80, 100 tons ha^{-1}) into soil 15 days before sowing had a superior effect on wheat than the application of NPK (40, 40, 10ha^{-1}) which indicated that the *Azolla* must have contributed to soil nutrient status. These studies all confirm that the addition of *A. filiculoides* improves soil fertility and leads to better yields showing that the nutrients released by the fern are availed to the plants for uptake and growth.

Previous studies show that *Azolla,* especially under lowland flooded conditions, has a positive effect on crop yields since it provides one of the most essential nutrients (N) needed in agriculture and enables uptake of other nutrients. In some cases the residual effect of the *Azolla* on rice has been utilised to increase yields in wheat. While a few studies have been done on the use of fresh *Azolla pinnata* as a biofertiliser of wheat, no comparison has been made on the short term and direct use of heated, fresh and dry *Azolla filiculoides* in dryland agriculture. It was hypothesised that *A. filiculoides,* a notorious water weed in South Africa, can be put to agricultural use under dryland conditions.

Therefore, in a preliminary study, the response of wheat to the application of the different treatments of *A. filiculoides* to sand was investigated under greenhouse conditions. Findings of the preliminary experiment allowed for the investigations to be carried out using *Azolla,* sand and topsoil in comparison to the performance of wheat fertilised with inorganic fertilisers. Most studies on the use of *Azolla* as a green manure have been carried out on rice under flooded conditions. In lowland rice cropping, Ito and Watanabe reported that the incorporation of *Azolla* in soil reduces the loss of N from the biofertiliser and enhances the availability of nutrients to plants. Rice grain and straw yields were improved by the addition of *Azolla*.

Several studies focused on the residual effect of the *Azolla* on wheat. The use of *Azolla* was found to have beneficial effects on grain and straw yields of rice with residual positive effect on the following wheat crop yields. Fewer studies have been done on the direct effects of *Azolla* on wheat. It has been found that the application of fresh fronds of *A. pinnata* harvested from a nearby pond increased the grain and straw yields of wheat. Pot grown *A. pinnata* that had been enriched with phosphate fertiliser was incorporated into soil (3 t ha^{-1}) to investigate the response of wheat.

The fresh fronds had better effect on wheat grain yield than the dry biomass, although both hardly affected the straw yields. But in this study, at the 20% and 80% biofertiliser rate, both total above ground biomass and grain weight per plant were significantly greater in the dry treatments than in the fresh treatments.

In most of the studies cited, emphasis was placed on the final grain and straw yields. However in this study, a wider range of growth parameters were measured at regular intervals throughout the growth period of the wheat plants. Furthermore, no previous study has taken the direct application of dried A. *filiculoides* and its effect on wheat as the first crop into account. This is important since the levels of N released may vary with *Azolla* species and their environment.

In addition, these studies have been carried out in Asia and not in South Africa where *A. filiculoides* has been found to grow profusely on water bodies and is a weed. At the commencement of this study, biological control of *A. filiculoides* had not been fully investigated and found successful. It was therefore hypothesised that the use of the weed as a biofertiliser would be a contribution to weed control on small water bodies such as farm dams and an effective use of a natural resource in South Africa.

NUTRIENTS AND PLANT BIOMASS

In a study on the release of N from nitrogenous plant materials, Muller and Sundman found that there was a high retention of plant derived N and relatively small loss of this N from the soil and that their value as supply to subsequent crops was long-term. This agrees with Ventura *et al.*, who found that the addition of *A. microphylla* significantly increased rice grain yields only after the second application of *Azolla* at the second crop harvest.

However, short term effects from the addition of the *Azolla* fern were observed for all treatments in both this preliminary and final study. This may be explained by earlier findings that the addition of *Azolla* increased nutrient availability as well as narrowing the C/N ratio in the soil that leads to faster decomposition of organic matter. In preliminary experiments of this study, the plants grown on treatments SF80, SD50 and SH80 produced the most dry matter. This would suggest that the pre-treatment and amount of the fern present in the soil affects nutrient release and availability in the soil.

Furthermore, the heated and dried *Azolla* effected significantly greater above ground biomass than the fresh *Azolla*. This difference in above ground biomass may be a result of nutrient availability due to the *Azolla* incorporated into soil as found by Watanabe and Ramirez, 1990 and Watanabe *et al.*, 1991. The beneficial results showed by application of dried *Azolla* grown plants are especially relevant as air drying reduces the bulk of the *Azolla* and makes it easier to transport.

The preliminary study agrees with previous studies in that the addition of *A. filiculoides* to acid washed sand improved the growth of wheat plants since the response of wheat in all the three treatments had significantly greater growth parameters than the pure sand control. This was confirmed in the main study. The sand was poor in nutrients due to acid washing and therefore good response of plants can be attributed to the organic matter and nutrients released into sand by the decomposition of the *Azolla* fern.

From this study, the addition of nitrogenous plant material to nutrient poor sandy soils resulted in increase yields over short term conditions and therefore one would expect beneficial long term results due to residual effects of the *Azolla*. Dry matter production in a plant is a result of the length of production period and the rate of dry matter accumulation in that period. A lack of nutrients in the soil may result in early maturation of the plant, although in this study, it only resulted in less accumulation of dry matter by SC grown plants.

Studies on the yield of mungbean and rice found yield increases with the addition of *Azolla* to the soil. In both the preliminary and final study, total above ground biomass yields were increased in all *Azolla* sand treatments. When the fertilisation rate was increased from 20% to 80%, the dry biomass of the wheat plants also increased. The preliminary study results showed that an increase (20% to 80%) in the amount of fresh and heated *Azolla* added per volume of sand had a significant effect, yet for the dry *Azolla*, the amount of fern added per volume of sand made no significant difference to the total above ground biomass.

In the main experiment, significant differences in total above ground biomass were observed at the 50, 75 and 128 day harvests between the 20% and 80% *Azolla* sand treatments. This difference in the dry biomass between the preliminary and final study may have been due to the sample size differences. It may also be due to the differences in the time of harvesting of *Azolla*. When topsoil was used, the effect of the fertilisers in topsoil on the total above ground dry biomass was greater than that in the control but the difference was lower than that observed in sand.

In sand, the straw biomass was greatly affected by fertilisation especially in the S80 treatment. In top soil, the fertilisers had no significant effects on the straw biomass. The use of *Azolla* in lowland agriculture has been compared to use of legumes as well as inorganic fertilisers. Watanabe & Liu compared the incorporation of Urea (57 N ha^{-1}), *Azolla* (84 kg N ha^{-1}) and legumes (73kg ha^{-1}) and found rice yields to be 6 tons ha^{-1}, 6.6 tons $ha^{-1}$and 6.1 tons ha^{-1} respectively. The application of *Azolla* increases the N, P and K content of the soil and leads to improved crop yields.

Lumpkin and Plucknett stated that 500 kg of fresh biomass gives 35-50 kg of dry biomass which contains N, P, K and organic matter which confirms the presence of N, P and K in *Azolla*. In this study, no direct measurements were made on the presence of nutrients in the soil and crop yields but the response of the wheat growth parameters was used as a measure of the soil fertility. Ram *et al.*, found that the addition of *Azolla* (24 tons ha^{-1}) to soil, increased ammonium and nitrate N resulting in optimum yields of pods, grain and plant weights of mungbeans (*Vigna radiata* L.).

In pot grown *Nardus stricta*, the addition of N, P and K (3:1:2) significantly increased above ground biomass. In the preliminary experiment, it was observed that the addition of fresh, dried and heated *A. filiculoides* to sand increased the grain weights, dry biomass, culm length and leaf areas of the wheat plants. In the main experiment, the addition of dried *A. filiculoides* increased wheat yields in sand but had less effect in topsoil.

Throughout the growth period, above ground biomass harvested from inorganically fertilised sand was as good as that in 20% *Azolla* treated sand. This is similar to a study by Galal who used isotope labelling to estimate efficiency of *Azolla* and urea as N sources and agrees that *Azolla pinnata* is equivalent to the use of urea in a clay loam soil under rice cultivation.

Furthermore, Talley and Rains incorporated 40 kg N ha^{-1} of dry *A. filiculoides* with spring rice in 1977 and obtained results equivalent to the use of ammonium sulphate, which supports the results in this study in which the biofertiliser increased dry biomass as much as the inorganic fertiliser or even better in nutrient poor soils.

In the main experiment, 80% *Azolla* in sand had the best effect and increased the above ground biomass by 220% and 342% at the 75 and 128 day harvests respectively in comparison to the control. Similar trends were observed in straw yields throughout the growth period. The use of *Azolla* as a biofertiliser has been found to be comparable to legume use due to its high N content.

The improved dry matter accumulation by wheat plants in comparison to the control in sand can be attributed to the increase in the amount of N available to the plants. For notably significant effects, greater amounts of the biofertiliser need to be added and this may be over a long term depending on the soil. This may explain why the wheat plants in all topsoil treatments did not show significant differences in the dry biomass accumulation at the 75 day harvest, as observed in the sand treatments.

Significant differences in above ground and straw biomass in topsoil treatments were observed at the 50 day harvest whereby the plants grown in the control and 20% *Azolla* had similar above ground biomass to those plants harvested from all the sand fertilised treatments. This would suggest a higher nutrient availability in the inorganic and 80% *Azolla* topsoil treatments with higher effect on straw biomass. However by the 128 day harvests, the best average plant above ground biomass harvested was from S80 and T20 respectively. In a perennial grass, *Molinia,* increase in N nutrition resulted in an increase of above ground productivity. In this study, this was notable in the sand treatments but not in the topsoil treatments.

Although, S80 plants flowered later than S20 plants, the S80 plants had the highest above ground biomass which is similar to the study on *Molinia* in which no detectable change in the rate of dry matter accumulation was found between the vegetative and reproductive stage.

The increased above ground dry biomass was a result of increases in the straw and grain due to N application especially in the sand-fertilised treatments. All plants in the topsoil treatments flowered at the same time and their final biomass does not differ as much as that of the sand treatments. This shows that flowering was not affected by the addition of fertilisers to topsoil.

Highest Grain Yield

Although the N content of the harvested plants was not tested in this study, the improved response of above ground matter and grain is assumed to be partly caused by the presence of N as a nutrient that plants take up after the decomposition of the biofertiliser or as supplied by the inorganic fertiliser. The highest yield (grain and plant height) in field grown winter wheat was recorded with 210 kg N ha^{-1} and N application affected the accumulation of biomass up to heading. In this study, the highest grain yield was obtained with the addition of 80% *Azolla* in sand and 20% *Azolla* in topsoil.

However, during the preliminary study, the highest grain yields were obtained in the 80% heated *Azolla* in sand. N uptake during grain filling was correlated to N applied to wheat which would suggest that in this study, the supply of N in fertilised topsoil was nearly as good as that in the topsoil control. This fact may be a result of sufficient nutrient status in the topsoil or poor nutrient release by dried *Azolla* in topsoil. N is a key element in achieving high yields in cereals since it is needed for the metabolic processes and its rate of uptake is determined by supply and demand.

Delogu *et al.*, state that soil N must be high at heading and grain filling for enhanced weight and high grain protein content. This may be the reason for a high average grain weight for the S80 grown plants. These plants flowered later and therefore more nutrients may have been available in the sand for both heading and post heading. Plants grown in SH80, S80 and T20 appear to have had sufficient nutrient supply during plant growth up to heading, especially N which has a dominant role in dry matter accumulation and N. Austin *et al.*, found that N present in plants at anthesis and at maturity was strongly correlated with total above ground weight at these times.

Therefore, the plants with the highest dry biomass would have the highest content of N which would be in the S80 and T20 grown plants in the main experiment. For a farmer, the grain is a very important part of the wheat plant, although the measurement of below ground matter is also useful. Spikelet number per ear and floret number per spikelet are important as it determines grain number and yield. Ewert and Honermeier investigated the duration of spikelet initiation and found that in winter wheat, it was not affected by N application (200 kg N ha^{-1}) but the number of spikelets per ear was increased.

This agrees with current studies whereby biofertilisation increased the grain yield per plant. Spikelet abortion was not prevented by N application a finding that may explain some results in this study on *Azolla* biofertiliser. However, in both the main and final experiments, the application of *Azolla*, led to an increase in the number of spikelets per ear relative to the controls. The application of N generally results in more grain production until some other factor, such as moisture or phosphate, becomes limiting. In this case, the limiting factor may be nutrients. Most of the N that the plant takes up before grain filling stage is used for the growth of non-reproductive structures, after which it contributes to grain production.

In winter wheat the concentration of grain N and shoot N decreased as the grain yield increased. The application of N fertiliser to winter wheat in Montana significantly increased the grain protein percentage and bread loaf volumes. MacDonald, found that grain weight and protein were affected by N application which is observable in the sand treatments but is not as clear in the topsoil treatments.

Therefore, in SC and topsoil treatments, grain differences may be due to lower supply of assimilate from source tissues or a slower rate of utilisation of available assimilate within the grain. The number of seeds or fruits and flower initiation can be affected by mineral nutrition which is clearly the case with various micronutrients.

In wheat, copper deficiency affects the reproductive phase as it inhibits anther formation, but hardly affects straw production. Therefore the poor grain formation in the SC plants, as compared to the fertilised plants, could have been a result of low micronutrient levels in the soil. Boron is essential for pollen tube formation and a low supply inhibits flowering and seed production. A limited supply of mineral nutrients available for re-translocation from source to sink may affect grain yield rather than a limited carbohydrate source.

The average grain size of a plant grown in sand treatments was not different from that of plants grown in topsoil. However, the number of grains per plant and therefore the total grain weight per plant differed between treatments. The increase in biofertiliser from 20% to 80% *Azolla* resulted in significant grain weight increases per plant. The SNPK grown plants had more grain than the S20 plants but less than the S80 plants which makes the S80 treatment better in sand than the inorganic fertiliser.

Topsoil grown plants had similar values of grain weight per plant and may therefore appear to have had sufficient micronutrient supply.

Relative Growth Rate (RGR)

As the plant grows the non-growing and non-photosynthetic material increases leading to a decline in relative growth rate which was observed in all treatments in the preliminary and final experiment. During the first 50 days, there is a high relative growth rate due to the addition of mass in form of growing points and photosynthetic area. Most of the dry matter was accumulated during this period. The period of dry matter accumulation was the same in all treatments without any differences in RGR between treatments over a particular time interval.

Gomez-Macpherson *et al.*, found no difference in RGR despite large differences in development rate in wheat lines within each isogenic set and photoperiod. Furthermore, the lack of difference in RGR calculated using above ground biomass was attributed to an increase in net assimilation rate in the early lines. No difference was found in isolines RGR despite phenological differences in spaced plants.

The lack of significant difference in relative growth despite differences in flowering and leaf areas may be due to a higher photosynthetic efficiency in the later flowering plants in the above ground biomass. However, in this study, the greater leaf area did not result in greater relative growth rate despite the increases in biomass due to fertilisation.

Leaf Area (LA)

In a study on winter wheat, the dry weights of leaves and stems at anthesis and maturity was increased by N, as was leaf area. In winter wheat, Austin *et al.* found a strong correlation between dry matter accumulation and N content which are affected by photosynthetic carbon fixation. N affects the dry matter since it has an effect on leaf area and therefore the response of plant leaf areas is important.

In the preliminary experiment, at 20% and 80%, the heated treatment effected a significantly higher leaf area on wheat plants and more above ground biomass accumulation. In the main experiment, all fertilised sand treatments effected significantly greater leaf areas on the wheat plants than the SC treatment with significantly greater above ground biomass accumulation. In topsoil, the T80 and TNPK effected significantly greater leaf areas on wheat plants but did not result in the most above ground biomass accumulation among the topsoil treatments.

Vouillot and Devienne-Barret found that N accumulation and translocation between the organs of vegetative winter wheat was affected by soil N availability. Furthermore under N deficiency, young leaves grew by utilising N remobilised from older leaves but under fertilised N conditions both soil and remobilised N contributed to leaf growth. This may explain the differences in leaf areas between the control plants and the fertilised plants in this study. This would mean that SH80 and S80 which had the highest leaf area, had a high supply of soil N and remobilised N.

Among the factors required for plant growth are mineral nutrients and an increase of these, will increase the growth rate to an optimum value. Supply at super optimal levels reverses increased productivity and mineral availability interacts with water availability. However throughout this study, water was not a constraint, and increased leaf areas that contributed to increased above ground biomass can be attributed to the availability of mineral nutrients. Langer grees that the addition of nutrients to cereal crops affects leaf area and numbers.

Investigations on a perennial grass (*Molinia)* showed that N allocation to leaves increased with N supply.

In wheat, Wojcieska found that the growth was strongly affected by the N nutrition in terms of dry weight and dry matter distribution with beneficial effect on the size of the leaves, a phenomenon that was observed in both the preliminary and final experiments on the biofertilisation with *Azolla*. The 20% *Azolla* treatments had similar effects on leaf areas of the sand and topsoil grown plants. This is in accord with a study by Baxter *et al.*, on montane grasses.

The plants harvested from SH80 and S80 had the largest leaf area and accumulated the greatest dry biomass. T80 grown plants had the greatest leaf areas among the topsoil treatments but T20 grown plants accumulated the most biomass with less green maximum leaf area. In both topsoil and sand treatments, plants fertilised with 80% and inorganic fertiliser had the greatest leaf areas. It is important to note that Gomez-Macpherson *et al.* found that mature leaf blade size was associated with delayed floral initiation.

Aerts found that on the whole, late developing wheat had the largest total leaf area per plant. Among the sand treatments in this study, the S20 grown plants flowered earliest and attained their maximum green leaf area earlier than the S80 and SNPK grown plants. However, fertilisation did not appear to affect the time of flowering and leaf areas in topsoil treatments.

Specific Leaf Area

Specific leaf area (SLA) as a measure of the accumulation of carbon per unit of green leaf area will decrease with time as the leaves mature, with lower values for the more efficient carbon-investing plants. Plants with higher SLA should have a higher RGR. In wheat, Gomez-Macpherson found that a decline in RGR was associated with a decline in LAR. This is similar to results from the preliminary study in which RGR and LAR declined over time while in the main study RGR and SLA showed the same trends in all treatments.

Aerts found that *Molinia,* a highly productive grass species, allocates most of its biomass to the roots but compensates the low biomass allocation to the leaves by a high SLA. In a study on herbaceous plants, Meziane and Shipley, found that SLA was affected by both irradiance and nutrient supply. This differs from the main experiment in which there were no observable significant differences in SLA in all treatments. Therefore the increases in leaf areas did not result in more efficient dry matter accumulation in either sand or topsoil, although differences were observed at the 50 and 75 day harvests.

Root Biomass

Prasad and Ram recorded increased dry root biomass of wheat plants at 62 days due to the application of *Azolla* to alluvial soil. Root biomass was found to increase with increase in *Azolla* application to a value of 100 tons ha^{-1}. In pot experiments, it was found that the addition of N, P and K fertiliser increased the root biomass of *Nardus stricta.* Findings by Aerts show that biomass allocation to roots in *Molinia,* a heathland grass species, was lower at high than at low N supply. In this study, increased root biomass in S20, S80 and SNPK could be the result of increased N supply in *Azolla* and inorganic fertiliser treatments.

Split pot experiments showed that at 80% *Azolla* biofertilisation rate, there is a saturation of nutrients that contributes to increased root branching and thus improves plant nutrient uptake. Sand was used because it is easier to harvest the roots although the presence of the *Azolla* made it difficult to remove all the roots and so the results may not be accurate due to loss of root material in the sand-*Azolla* treatments. In the split pot experiments, significantly more roots were collected from the *Azolla* side than from the pure acid washed sand.

The root biomass collected from the sand treatments in the main experiment showed more root biomass in the inorganic and *Azolla* fertiliser treatments than in the control. In the preliminary experiment, R:S ratios of SC plants were higher than the R:S ratios of plants harvested from dried *Azolla* treatments. In the main experiment, R:S ratios were highest in the SC treatment suggesting a higher investment in their underground biomass than in the above ground biomass.

Soil Moisture

Due to the high amount of organic matter added to a soil by the incorporation of *Azolla*, soil fertility was improved. Ram *et al.* found that the addition of *Azolla* to a sandy loam soil (pH 7.8) improved soil physical properties and water holding capacity. In a long term rice-wheat cropping, green manure *Azolla* proved to be as good as inorganic fertiliser and improved soil structure. This was observed in the sand treatments in this study. Mubarik found that *Azolla* increased the water holding capacity of soil. In both the preliminary and final study, the incorporation of *A. filiculoides* in sand supplied nutrients and increased the water holding capacity of sand.

The best soil-water content was observed in the preliminary study in the SH80, SF50 and SF80 *Azolla* treatments. This would be an important factor under field conditions for nutrient poor sandy soils. Topsoil treatments had different results from those observed in the sand treatments. There were no significant effects on soil water content between the control and fertilised the topsoil treatments.

The soil water content in the 20% and 80% *Azolla* sand treatments was as good as that in the topsoil treatments.

Previous studies and this study show that *A. filiculoides* increases above ground biomass, root ramification, grain weight, leaf area in crops. As an organic fertiliser, it also improves soil properties. Although it has mainly been applied as a green manure, Ram and Prasad, and results from this study do confirm that *Azolla* can be used under dryland conditions to compliment and substitute inorganic fertilisers.

Possible Problems

Spore formation

Spore formation by *Azolla* has been found to be associated with mat senescence. Ashton determined that *A. filiculoides* spores take a period of 17 to 43 days to germinate in the laboratory after the artificial combination of megaspores and microspores. Janes found that in Britain, at 20°C, *A. filiculoides* spore germination took 11-25 days while at 15°C, it took 40-63 days. During the main experiment, unlike in the preliminary one, the spores germinated following extraction from both sand and soil treatments.

The spores in sand took 74 days to germinate, while those in topsoil took 96 days to germinate. In the preliminary study, spore germination may not have occured due to absence of spores in the *Azolla* used as a result of the time of harvest. Harvesting might be ideal during spring or early summer before the multi-layered mats are formed by the *Azolla* plants.

When harvesting, care must be taken to choose mats that are green since the red mats may contain spores and lower concentrations of N and P concentrations. In the preliminary study, the length of time may not have been enough for spores to germinate. This would suggest that the spread of *Azolla* when used in biofertilisation is very unlikely. In view of the germination of spores during the main experiment, a problem arises in the use of *A. filiculoides* as a biofertiliser.

Heavy metal release into soil

Several studies have showed that *Azolla* takes up heavy metals especially from polluted waters. *Azolla* as fresh and dry matter is known to take up heavy metals such as Zn, Cu, Cr, U and Hg. This would affect the use of *Azolla* as a biofertiliser since plants would take up the metals if made available in the soil water. It would necessitate the testing for 0metals in the *Azolla* before application into soil. However the uptake of metals like mercury is dependant on their concentration in water.

The uptake of metals was found to affect the health of the fronds making their lifespan very short and causing them to change colour from green to red. Furthermore, the presence of metals in *Azolla* decreased the nutrient (Mg & K) value of the plants. *Azolla* was found to be more resistant to iron and copper uptake than *Lemna minor* L. Therefore release of metals into soil by *Azolla* and uptake of metals by crops needs to be investigated before large scale use of *Azolla* as a biofertiliser when metal pollutants are likely to be present. The source of the *Azolla* is thus important and must not be from water bodies in industrial areas.

REFERENCES

Agren, G. I., "Theory for growth of plants derived from the nitrogen productivity concept," *Physiol. Plant*, 64, 17-28, 1985.

Ashton, P. J. and Walmsley, R. D., "The aquatic fern *Azolla* and its *Anabaena* symbiont," *Endeavour*, 1976, 35, 39-43.

Baxter, R., Ashenden, T., W. Sparks, T. H. and Farrar, H., "Effects of elevated carbon dioxide on three montane grass species 1, Growth and dry matter partitioning, *J. Expt. Bot.* 45, (272) 305-315, 1994.

Ewert, F. and Honermeier, B., "Spikelet initiation of winter *Triticale* and winter wheat in response to nitrogen fertilisation," *European J. Agronomy* 11;107-113, 1999.

Garnier, E., "Growth analysis of congeneric annual and perrenial grass species," *J. Ecology*, 80, 665-675, 1992.

Hill, M. P., "*Azolla filiculoides* the first step towards biological control," *Plant protection unit*, 51, 1-3, 1998.

Ito, O and Watanabe, I., "Availability to rice plants of nitrogen fixed by *Azolla*," *Soil Sci. Plant Nutr.*, 31 (1) 91-104, 1985.

Janes, R., "Growth and survival of *Azolla filiculoides* in Britain I. vegetative reproduction," *New Phytol.* 138, 367-375, 1998a.

Kondo, M, Kobayashi, M and Takahashi, E., "Effect of *Azolla* and its utilisation in rice culture in Niger," *Plant and Soil* 120, 165-170.

Ram, H. and Prasad, J., "Comparative study of *Azolla* and N, P, K fertilisers on wheat crop," *Agric. Res. Rural Dev.* 5, 43-45.

9

Microbial Phosphate Solubilisation

Phosphorus is second only to nitrogen in mineral nutrients most commonly limiting the growth of crops. Phosphorus is an essential element for plant development and growth making up about 0.2 % of plant dry weight. Plants acquire P from soil solution as phosphate anions. However, phosphate anions are extremely reactive and may be immobilised through precipitation with cations such as Ca^{2+}, Mg^{2+}, Fe^{3+} and Al^{3+}, depending on the particular properties of a soil.

In these forms, P is highly insoluble and unavailable to plants. As the results, the amount available to plants is usually a small proportion of this total. Several scientists have reported the ability of different bacterial species to solubilise insoluble inorganic phosphate compounds, such as tricalcium phosphate, dicalcium phosphate, hydroxyapatite, and rock phosphate.

PHOSPHATE SOLUBILISATION

The principal mechanism for mineral phosphate solubilisation is the production of organic acids, and acid phosphatases play a major role in the mineralisation of organic phosphorus in soil. It is generally accepted that the major mechanism of mineral phosphate solubilisation is the action of organic acids synthesized by soil microorganisms.

Production of organic acids results in acidification of the microbial cell and its surroundings.

The production of organic acids by phosphate solubilising bacteria has been well documented. Gluconic acid seems to be the most frequent agent of mineral phosphate solubilisation. Also, 2-ketogluconic acid is another organic acid identified in strains with phosphate solubilising ability (Table 1).

Table 1. Microbial strains producing organic acid

Organic acid	*Strains*
Gluconic acid	*Pseudomonas* sp., *Erwinia herbicola, Pseudomonas cepacia,* Burkholderia cepacia
2-Ketogluconic acid	*Rhizobium leguminosarum, Rhizobium meliloti, Bacillus firmus*

Strains of *Bacillus* were found to produce mixtures of lactic, isovaleric, isobutyric and acetic acids. Other organic acids, such as glycolic, oxalic, malonic, and succinic acid, have also been identified among phosphate solubilisers. Strains from the genera *Pseudomonas, Bacillus* and *Rhizobium* are among the most powerful phosphate solubilisers.

Chelating substances and inorganic acids such as sulphideric, nitric, and carbonic acid are considered as other mechanisms for phosphate solubilisation. However the effectiveness and their contribution to P release in soils seems to be less than organic acid production.

Mineral Phosphate Solubiliser

Detection and estimation of the phosphate solublisation ability of microorganisms have been possible using plate screening methods. Phosphate solubilisers produce clearing zones around the microbial colonies in media. Insoluble mineral phosphates such as tricalcium phosphate or hydroxyapatite are contained in the media.

Also the bromophenol blue method that produce yellow halos following pH drop through the release of organic acids is more reproducible and has greater correlation in comparison with the simple halo method. However, clearing zones on agar plate method is generally used (Figure 1). Pikovskàys's medium (Table 2) is a general medium for selection of phosphate solubiliser.

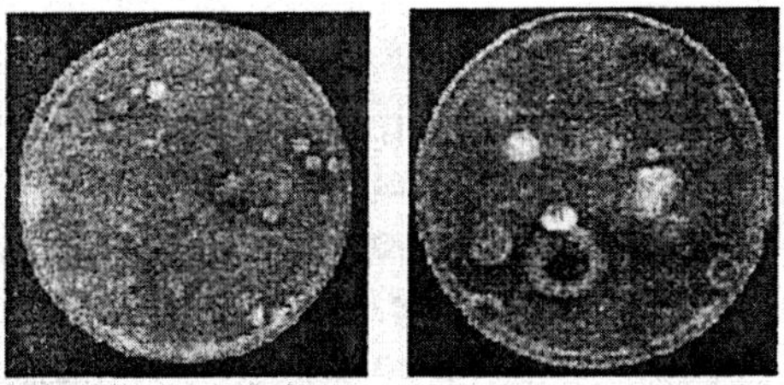

Figure 1: Phosphate solubiliser forming clear zone. Bacteria (left), Fungi (right)

Table 2. Composition of medium for phosphate solubilisers (Pikovskays's medium)

Components	*Amounts* ($g\ l^{-1}$)
Glucose	10
$Ca_3(PO_4)_2$	5
$(NH_4)_2SO_4$	0.5
NaCl	0.2
$MgSO_4.7H_2O$	0.1
KCl	0.2
Yeast extract	0.5
$MnSO_4.\ H_2O$	0.002
$FeSO_4.7H_2O$	0.002
pH	7.0

Production of Inoculant

Microorganisms with potential as plant growth promoters have been used to produce inoculants. Potential materials that are able to support good growth and survival of bacteria are needed in inoculant production. Many materials have been evaluated, including different types of coals, bentonite, corn oil, mineral soils, peat, peat moss, vermiculite, and perlite. Peat is commonly used material for inoculant carrier. Finely ground peat is most commonly used in conventional inoculant production.

However, peat is not always available, because some peat may inhibit growth of some *Rhizobium* strains. There have been reports on the difficulty of obtaining autoclaved or gamma-irradiated peat as carriers. The high temperature during steam sterilisation or the high dosage needed for irradiation might generate toxic substances for bacteria. However, perlite can be easily sterilised with no risk of producing toxic substances, because it is a volcanic stone, composed of a little-hydrated aluminium silicate. The agronomic use of agro-wastes as substrates causes changes in the soil affecting its physico-chemical characteristics and microbial activity in the rhizosphere.

The breakdown of such materials to simple sugars provides energy sources for heterotrophic microorganisms such as P-solubilising and nitrogen fixing bacteria. Normally, the growth and metabolic activity of soil microorganisms are limited by the availability of nutrients. From this reasons, several kinds of agro wastes such as rice straw compost will be good carrier materials for the inoculants and improver of soil condition as following case study. However, the peat moss with phosphate solubilizer showed good survival of inoculants and effects on crops.

Cultivation

— *Media:* proper media for inoculant (nutrient broth, yeast extract broth etc.)
— *Incubation condition:* temperature, light, incubation period.

Collection

- After incubation period, collect microbial cells by centrifuge
- Wash cells with distilled water or diluted saline solution.

Applications of Inoculant

Inoculation method of phosphate solubilizer

Generally biofertilisers in powder form are applied as for organic matters onto the soil. This type is very convenient for users on the management of biofertiliser. Some biofertilisers are costly products for farmers, so their use would be restricted on the specific condition of agronomy.

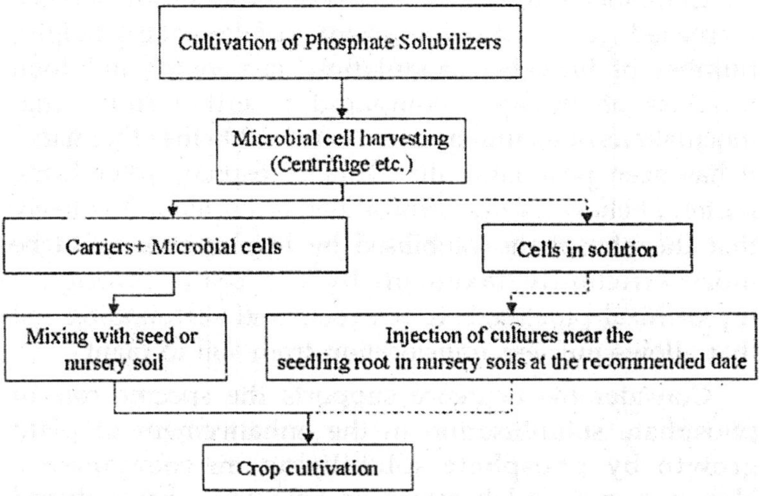

Figure 2: Inoculation method of phosphate solubilizers.

Microorganisms are generally supplied by producers of biofertiliser, so it would only necessary that the users or farmers follow the application method prepared by manufacturers. However, the popular application method is regarded as next procedure (Figure 2).

Improvements of phosphate solubilisers

An alternative approach for the use of phosphate solubilising bacteria as microbial inoculants is the use of mixed cultures or co-inoculation with other microorganisms. This evidence points to the advantage of the mixed inoculations of PGPR strains comprising phosphate solubilising bacteria.

The effect of a combined inoculation of *Rhizobium*, a phosphate solubilising *Bacillus megaterium* sub sp. *phospaticum* strain-PB and a biocontrol fungus *Trichoderma* spp. On growth, nutrient uptake and yield of chickpea were studied under glasshouse and field conditions.

Combined inoculation of these three organisms showed increased germination, nutrient uptake, plant height, number of branches, nodulation, pea yield, and total biomass of chickpea compared to either individual inoculations or an uninoculated control. On the other hand, it has been postulated that some phosphate solubilising bacteria behave as mycorrhizal helper bacteria. It is likely that the phosphate solubilised by the bacteria could be more efficiently taken up by the plant through a mycorrhizal pipeline between roots and surrounding soil that allows nutrient translocation from soil to plant.

Considerable evidence supports the specific role of phosphate solubilisation in the enhancement of plant growth by phosphate solubilising microorganisms. However, not all laboratory or field trials have offered positive results. Therefore, the efficiency of the inoculation varies with the soil type, specific cultivars, and other parameters.

Improvement of soil condition for phosphate solubilisers

A considerably higher concentration of phosphate solubilising bacteria is commonly found in the rhizosphere soil. Also the fungal genera with this capacity are *Penicillium* and *Aspergillu*.

A continued exploration of the natural biodiversity of soil microorganisms and the optimisation and manipulation of microbial interactions in the rhisosphere of crops represents a prerequisite step to develop more efficient microbial inoculants. Also, it is desirable to find methods to manage indigenous microbes in soil for sustainable agriculture. The answer may be obtained from diverse field tests managed for improvement of soil condition. A case study for phosphate solubilisers has been conducted on a fifty-year long-term experiment in rice paddy soils in Korea.

An investigation was made on the effect of different soil amelioration on the physical and chemical properties of soil, the yield or quality of rice, and the activity of soil microbes in the plot where NPK, lime, and compost had been applied for fifty years. The physical properties were most significantly affected by the application of compost, which reduced the bulk density and hardness of soil and increased the liquid and gas phase of soil and cation exchange capacity of soil.

The application of compost tended to increase organic matter, available phosphorus and silicate, exchangeable cations in the soil.

Microbial diversity of soils

Diversity is an important concept in ecology, often in environmental monitoring and conservation management. But diversity is an intuitively obvious concept that is difficult to define and quantify. There are two parts to diversity: the number of species and the evenness of the distribution. A sample with more species is more diverse: while an evenly distributed community is more diverse than an unevenly distributed community with the same number of species.

However, microbial communities are severely affected by agricultural managements such as tillage, and organic materials and chemicals for soil amendment. A pattern was revealed from the long-term experiment showing different profiles of the ratio of functional species to bacteria. The ratios of fluorescent bacteria, Bacillus and Gram-negative bacteria to total aerobic bacteria in CLAPK (compost + lime + fertiliser) and CAPK (compost + fertiliser) plots were higher than other treatments. This result represents effect of amendments on soil microbial community.

Nevertheless microbial diversity was simplified, the functional groups were triggered by the application of substances to soil. Apart from fertilisation and enzymatic decomposition of organic compounds, microbial P-mobilisation would be the only possible way to increase plant-available phosphorus. Many species of bacteria are able to solubilise phosphates in vitro and some of them can mobilise P in plants.

Table 3. Effect of materials on available phosphorus and PB/TB ratio in paddy soils

Treatment	*PH (1:5)*	*Av.*P_2O_5 *(..-1)*	*PB/TB ratio (%)*
Control	6.0	40	0.6
F	5.6	73	2.2
FL	6.3	73	5.0
FC	5.6	87	6.3
FLC	6.1	105	16.8

The phosphate solubilising activity is significantly affected by material input. High positive correlation was observed between the phosphate solubilising bacteria and organic matter contents in soils. The treatment of the agricultural materials is presented in Table 3. The available phosphorus contents ranged from 40 to 105 mg. FLC contained highest amount of available phosphorus than control. FLC showed high ratio of PB/TB between treatments. This means that lime and compost used as soil improver had positive effects on phosphate solubilisers.

Modern agricultural practices largely rely on high inputs of mineral fertilisers to achieve high yield. These practices are now being reevaluated and are coming under increased scrutiny as our awareness of environmental conservation from the impact of chemical overuse. This research is very necessary to improve soil condition on environmentally friendly agriculture that is being practiced at present.

Table 4. Effects of amendment on phosphate solubiliser in soils

Treatment	*Species of isolate*	*Number of isolate*	
	Bacillus simplex	1	
Chemical Fertiliser	Bacillus megaterium	2	4
	Pseudomonas chlororaphis	1	
	Bacillus pumilis	2	
	Pseudomonas chlororaphis	1	
Lime + fertiliser	Pseudomonas coronafaciens	1	5
	Arthrobacter globiformis	1	
	Bacillus pumilis	3	
	Bacillus licheniformis	2	
	Pseudomonas fluorescens	4	
	Bacillus marinus	1	13
	Bacillus subtilis	1	
	Bacillus lentimorbus	1	
Compost + fertiliser	Pseudomonas chlororaphis	1	

Also we should imagine that improvement of soil condition includes activation of functional microbes related to specific materials input. As shown in Table 4, diversity of phosphate solubilisers in soils was highly controlled by the agricultural management. Only three types of treatment are shown in the table. But as was commented earlier, compost directly affected microbial community in soils.

Effects of organic matters

Organic matter results in several benefits such as better soil structure that provides a more suitable medium for plant growth, supplies nutrients for the plant and also help to build up target microorganisms. The combined use of organic matter with inoculant in the build up of high bacterial populations that improve plant growth.

Table 5. Effect of materials on amount of nutrient absorbed by rice cultivated in paddy soils (unit: kg ha^{-1})

Treatment	*T-N*	*P_2O_5*	*K_2O*	*CaO*	*MgO*
Control	46.4(50)	33.2(60)	76.4(63)	22.9	12.0
NPK	93.4(100)	54.9(100)	120.4(100)	38.0	23.4
NPK+C	107.5(115)	64.9(118)	161.9(134)	41.9	28.6
NPK+L	89.7(96)	52.5(96)	124.3(103)	36.7	22.9
NPK+S	94.0(104)	53.8(98)	137.1(114)	37.7	25.0
NPK+C+L+S	106.3(114)	63.1(115)	164.0(136)	35.4	26.0

Effect of materials on amount of nutrient absorbed by rice is presented in Table 5. It is possible to suggest that the only driving force to dissolve phosphate is microbial activity, especially phosphate solubilising bacteria, from the results shown in the table.

REFERENCES

Rengel, Z. G. D. Batten, and D. E. Crowley., "Agronomic approaches for improving the micronutrient density in edible portions of field crops," *Field crops research* 60: 27-40, 1999.

Shenker, M., I. Oliver, M. Helmann, and Y. Hadar., "Utilization by tomatoes of iron mediated by a siderophore produced by Rhizopus arrhizus," *J. Plant Nutr.* 15, 2173-2182, 1992.

10

Plant Growth Promoting Rhizobacteria

Plant growth-promoting rhizobacteria (PGPR) were first defined by Kloepper and Schroth to describe soil bacteria that colonise the roots of plants following inoculation onto seed and that enhance plant growth. The following are implicit in the colonisation process: ability to survive inoculation onto seed, to multiply in the spermosphere (region surrounding the seed) in response to seed exudates, to attach to the root surface, and to colonise the developing root system.

The ineffectiveness of PGPR in the field has often been attributed to their inability to colonise plant roots. A variety of bacterial traits and specific genes contribute to this process, but only a few have been identified. These include motility, chemotaxis to seed and root exudates, production of pili or fimbriae, production of specific cell surface components, ability to use specific components of root exudates, protein secretion, and quorum sensing. The generation of mutants altered in expression of these traits is aiding our understanding of the precise role each one plays in the colonisation process.

Progress in the identification of new, previously uncharacterised genes is being made using nonbiased screening strategies that rely on gene fusion technologies. These strategies employ reporter transposons and *in vitro* expression technology (IVET) to detect genes expressed during colonisation. Using molecular markers such as green fluorescent protein or fluorescent antibodies it is possible to monitor the location of individual rhizobacteria on the root using confocal laser scanning microscopy. This approach has also been combined with an rRNA-targeting probe to monitor the metabolic activity of a rhizobacterial strain in the rhizosphere and showed that bacteria located at the root tip were most active.

An important aspect of colonization is the ability to compete with indigenous microorganisms already present in the soil and rhizosphere of the developing plant. Our understanding of the factors involved in these interactions has been hindered by our inability to culture and characterise diverse members of the rhizosphere community and to determine how that community varies with plant species, plant age, location on the root, and soil properties. Phenotypic and genotypic approaches are now available to characterise rhizobacterial community structure.

Phenotypic methods that rely on the ability to culture microorganisms include standard plating methods on selective media, community level physiological profiles (CLPP) using the BIOLOG system, phospholipid fatty acid (PLFA), and fatty acid methyl ester (FAME) profiling.

Culture-independent molecular techniques are based on direct extraction of DNA from soil and 16S-rRNA gene sequence analysis, bacterial artificial chromosome or expression cloning systems. These are providing new insight into the diversity of rhizosphere microbial communities, the heterogeneity of the root environment, and the importance of environmental and biological factors in determining community structure. These approaches can also be used to determine the impact of inoculation of plant growth-promoting rhizobacteria on the rhizosphere community.

MECHANISMS

PGPR enhance plant growth by direct and indirect means, but the specific mechanisms involved have not all been well-characterised. Direct mechanisms of plant growth promotion by PGPR can be demonstrated in the absence of plant pathogens or other rhizosphere microorganisms, while indirect mechanisms involve the ability of PGPR to reduce the deleterious effects of plant pathogens on crop yield. PGPR have been reported to directly enhance plant growth by a variety of mechanisms: fixation of atmospheric nitrogen that is transferred to the plant, production of siderophores that chelate iron and make it available to the plant root, solubilisation of minerals such as phosphorus, and synthesis of phytohormones.

Direct enhancement of mineral uptake due to increases in specific ion fluxes at the root surface in the presence of PGPR has also been reported. PGPR strains may use one or more of these mechanisms in the rhizosphere. Molecular approaches using microbial and plant mutants altered in their ability to synthesize or respond to specific phytohormones have increased our understanding of the role of phytohormone synthesis as a direct mechanism of plant growth enhancement by PGPR. PGPR that synthesize auxins and cytokinins or that interfere with plant ethylene synthesis have been identified.

PGPR that indirectly enhance plant growth via suppression of phytopathogens do so by a variety of mechanisms. These include the ability to produce siderophores that chelate iron, making it unavailable to pathogens; the ability to synthesize anti-fungal metabolites such as antibiotics, fungal cell wall-lysing enzymes, or hydrogen cyanide, which suppress the growth of fungal pathogens; the ability to successfully compete with pathogens for nutrients or specific niches on the root; and the ability to induce systemic resistance. Biochemical and molecular approaches are providing new insight into the genetic basis of these traits, the biosynthetic pathways involved, their regulation, and importance for biological control in laboratory and field studies.

CHALLENGES IN DEVELOPING PGPR

One of the challenges in developing PGPR for commercial application is ensuring that an effective selection and screening procedure is in place, so that the most promising organisms are identified and brought forward. In the agricultural chemical industry, thousands of prospective compounds are screened annually in efficient high-throughput assays to select the best one or two compounds for further development. Similar approaches are not yet in place for PGPR. Effective strategies for initial selection and screening of rhizobacterial isolates are required. It may be important to consider host plant specificity or adaptation to a particular soil, climatic conditions or pathogen in selecting the isolation conditions, and screening assays.

The spermosphere model, an enrichment technique that relies on seed exudates as the nutrient source, has been used for selection and isolation of promising N_2-fixing rhizosphere bacteria from rice. One approach for selection of organisms with the potential to control soil-borne phytopathogens is to isolate from soils that are suppressive to that pathogen. Other approaches involve selection based on traits known to be associated with PGPR such as root colonisation, 1-aminocyclopropane-1-carboxylate (ACC) deaminase activity, and antibiotic and siderophore production. The development of high throughput assay systems and effective bioassays will facilitate selection of superior strains.

Application Challenges

The application of PGPR for control of fungal pathogens in greenhouse systems shows considerable promise, due in part to the consistent environmental conditions and high incidence of fungal disease in greenhouses. Achieving consistent performance in the field where there is heterogeneity of abiotic and biotic factors and competition with indigenous organisms is more difficult. Knowledge of these factors can aid in determination of optimal concentration, timing and placement of inoculant, and of soil and crop management strategies to enhance survival and proliferation of the inoculant.

The concept of engineering or managing the rhizosphere to enhance PGPR function by manipulation of the host plant, substrates for PGPR, or through agronomic practices, is gaining increasing attention. Development of better formulations to ensure survival and activity in the field and compatibility with chemical and biological seed treatments is another area of focus; approaches include optimisation of growth conditions prior to formulation and development of improved carriers and application technology.

Commercialisation Challenges

Prior to registration and commercialisation of PGPR products, a number of hurdles must be overcome. These include scale up and production of the organism under commercial fermentation conditions while maintaining quality, stability, and efficacy of the product. Formulation development must consider factors such as shelf life, compatibility with current application practices, cost, and ease of application.

Health and safety testing may be required to address such issues as non-target effects on other organisms including toxigenicity, allergenicity and pathogenicity, persistence in the environment, and potential for horizontal gene transfer. The product claim, whether as a fertiliser supplement or for biological control, will determine to which federal agency applications for registration should be addressed in Canada and the USA. Capitalisation costs and potential markets must be considered in the decision to commercialise. McSpadden Gardener and Fravel estimated that a minimum capitalisation of $1 million US is required to register a biopesticide product in North America.

FUTURE EXPECTATION

As our understanding of the complex environment of the rhizosphere, of the mechanisms of action of PGPR, and of the practical aspects of inoculant formulation and delivery increases, we can expect to see new PGPR products becoming available. The success of these products will depend on our ability to manage the rhizosphere to enhance survival and competitiveness of these beneficial microorganisms. Rhizosphere management will require consideration of soil and crop cultural practices as well as inoculant formulation and delivery.

Genetic enhancement of PGPR strains to enhance colonisation and effectiveness may involve addition of one or more traits associated with plant growth promotion. Genetic manipulation of host crops for root-associated traits to enhance establishment and proliferation of beneficial microorganisms is being pursued. However, regulatory issues and public acceptance of genetically engineered organisms may delay their commercialisation.

The use of multi-strain inocula of PGPR with known functions is of interest as these formulations may increase consistency in the field. They offer the potential to address multiple modes of action, multiple pathogens, and temporal or spatial variability. PGPR offer an environmentally sustainable approach to increase crop production and health. The application of molecular tools is enhancing our ability to understand and manage the rhizosphere and will lead to new products with improved effectiveness.

PGPR and Biodegradation

Some soil bacteria preferentially associate with the roots of crop plants and can exert beneficial effects on their hosts. They are collectively referred to as PGPR (Plant Growth Promoting Rhizobacteria). PGPR strains are often found among the fluorescent pseudomonads. Biocontrol of phytopathogens appears to be a major mechanism of plant growth promotion by these bacteria. Suppression of phytopathogens results from the production of secondary metabolites or by elicitation of the plant's own defence system.

PGPR-based inocula must be able to compete with the indigenous micro-organisms and efficiently colonise the rhizosphere of the plants to be protected. Previously, we investigated the role of *Pseudomonas fluorescens* cell envelope proteins in adhesion to roots. In addition, it was demonstrated that chemotaxis for root exudates is a major determinant for competitive root colonisation. Split-screen movies showing the swimming behavior of wild type versus *cheA* mutant can be viewed for *P. fluorescens* strains F113, OE28.3, SBW25, and WCS365. Current work focuses on the contribution of bacteriocin production to rhizosphere competence of *fluorescent Pseudomonas.*

Bacteriocins

Bacteriocins are peptides or proteins that selectively kill related bacteria (of the same species or genus) but do not affect other organisms. Bacteriocin-mediated antagonism is believed to occur in virtually any niche colonised by bacteria. Bacteriocin production is detectable is most strains of the human opportunistic pathogen *Pseudomonas aeruginosa* and has been used to differentiate clinical isolates. Only few of these bacteriocins, called pyocins, and their genes have been characterised at the molecular level. Likewise, bacteriocin production is ubiquitous among plant-associated *Pseudomonas.*

A novel type of bacteriocin (LlpA) was characterised in a *Pseudomonas* isolate from banana rhizosphere. Remarkably, LlpA shows similarity with a group of mannose-binding lectins that were originally discovered in plants but have a much wider distribution, including fungi and animals. Apparently, LlpA constitutes the prototype of a new family of antagonistic proteins. Biofertilisation represents another mechanism used by PGPR to enhance plant growth. The possibility of providing biologically fixed nitrogen to rice plants is explored for an endophytic nitrogen-fixing pseudomonad.

Interaction between Pseudomonas and rice

Rice (*Oryza sativa* L.) is the most important food crop currently produced and requires extensive use of chemical nitrogen fertilisers. Part of the required nitrogen may be provided by rice-associated or endophytic bacteria that are capable of biological nitrogen fixation, such as strain A15, isolated from rice in China. Although the ability to fix nitrogen by *Pseudomonas* species has long been debated, we unequivocally identified strain A15 as a diazotrophic *Pseudomonas stutzeri*.

To study the mechanisms that enable strain A15 to colonise and infect rice roots, and to survive within rice plants, a *dapB*-based 'in vivo expression technology' (IVET) strategy is used. IVET, originally devised for pathogenic bacteria, is a 'promoter-trapping' technique that traces bacterial promoters driving expression of an essential gene, specifically during interaction with the host.

BIODEGRADATION

In recent years there has been an increasing interest in the ability of bacteria to degrade various man-made chemicals and their use for bioremediation of polluted environments. The biochemical pathways involved are best understood for various Gram-negative bacteria. Several Gram-positive bacteria, and actinomycetes in particular, are also capable of degrading many recalcitrant compounds. However, appropriate genetic tools for these bacteria are often lacking. We have developed the *E. coli-Rhodococcus* shuttle vector pFAJ2574 based on a small cryptic *Rhodococcus* plasmid. Also, a transposon was constructed based on the indigenous *Rhodococcus* mobile element IS*1415*, a member of the IS*21* family.

As a model system for the breakdown of pesticides by actinomycetes, the biodegradation of thiocarbamate herbicides by *Rhodococcus erythropolis* NI86/21 was studied. Among the enzymes identified are a dedicated cytochrome P450 system (CYP116) and a non-heme haloperoxidase (ThcF). Recently, a number of CYP116-related fusion proteins were identified in different bacterial species. These so-called 'P450 PFOR' enzymes constitute a novel class of self-sufficient monooxygenases. ThcF belongs tot a large subfamily of a/b hydrolases for which no natural substrates have been identified yet. Expression of ThcF involves ThcG, a regulatory protein belonging to the recently defined LuxR subfamily 'LAL'. The hallmark of these unusually large regulators (> 800 amino acids) is the presence of an aminoterminal ATP-binding site in addition to the carboxyterminal DNA-binding site.

Proteasome of Actinomycetes

The proteasome constitutes a molecular machine that enables controlled ATP-dependent breakdown of intracellular proteins. In eukaryotes, the 26S proteasome is built of a 20S proteolytic core complex and one or two 19S regulatory complexes, containing AAA ATPases, attached to one or both ends of the cylinder-shaped core complex. Although other self-compartmentalising proteases occur in *Bacteria*, remarkably, genuine 20S proteasomes are only present in *Archaea* and one phylum of *Bacteria*, the actinomycetes.

GENETIC DIVERSITY AND SCREENING STRATEGIES OF PGPR

Rhizosphere Microbial Ecology

The German agronomist Hiltner first defined the rhizosphere, in 1904, as the effect of the roots of legumes on the surrounding soil, in terms of higher microbial activity because of the organic matter released by the roots. Until the end of the twentieth century, this effect was not considered to be an ecosystem. It is interesting to make some brief observations about the size, in terms of energy and extension, of this ecosystem to determine its impact on how the biosphere functions. First, in extension, the rhizosphere is the largest ecosystem on earth. Second, the energy flux in this system is enormous.

A large number of macroscopic organisms and microorganisms such as bacteria, fungi, protozoa and algae coexist in the rhizosphere. Bacteria are the most abundant among them. Plants select those bacteria contributing most to their fitness by releasing organic compounds through exudates, creating a very selective environment where diversity is low. A complex web of interactions takes place among them, and this may affect plant growth, directly or indirectly.

Since bacteria are the most abundant microorganisms in the rhizosphere, it is highly probable that they influence the plants physiology to a greater extent, especially considering their competitiveness in root colonisation. Bacterial diversity can be defined in terms of taxonomic, genetic and functional diversity. In the rhizosphere, the metabolic versatility of a bacterial population (functional diversity) is based on its genetic variability and on possible interactions with other prokaryotic and eukaryotic organisms such as plants. However, a question still to be answered regarding microbial communities in the rhizosphere is the relationship between the ecological function of communities and soil biodiversity.

In spite of the lack of information about the importance of the diversity and the richness of species related to their ecological function, soil organisms have been classified several times in functional groups. This lack of knowledge about bacterial diversity is partly owing to the high number of species present, as well as to the fact that most bacteria are viable but not culturable.

The biological diversity of soil microorganisms has been expressed using a variety of indexes and mathematical models, but there is no accepted general model to describe the relationship among abundance, species richness and dominancy. It is, therefore, reasonable that the components of diversity are studied separately to quantify them.

Bacterial diversity studies are more complex at taxonomic, functional and genetic levels than are similar studies on eukaryotic organisms owing to the minute working scale and the large number of different bacterial species present in the environment. Torsvik and coworkers identified more than 7000 species in an organic forest soil. The variations in populations through space and time and their specialisation in ecological niches are two important factors in the rhizosphere that must be considered in studying how species richness influences the functioning of the system.

The functioning of soil microbial communities is based on the fact that there is appropriate species diversity for the resources to be used efficiently and that this can be maintained under changing conditions. In the rhizosphere, as in other well-formed ecosystems with an appropriate structure, changes in some of the components can affect entire or part of the system. The degree of impact will depend on features of the system such as its resistance or resilience. The state of this system changes depending on variables such as the age of the plant, root area, light availability, humidity, temperature and plant nutrition.

Under stressful conditions, the plant exerts a stronger control on release of root exudates. From this viewpoint, it is reasonable to assume that the changes that occur in the plant will change the root exudation patterns and, thereby, the rhizosphere microbial communities. There have been many studies that relate the quality and quantity of the exudates with changes in the structure of rhizosphere microbial communities. In 1980, Torsvik published the first protocol for the extraction and isolation of microbial DNA from soil. Since then, there have been many studies directed at the development of new methods and molecular tools for the analysis of soil microbial communities. However, molecular genetics is not the only tool used in solving the difficulties in analysing soil microbial communities.

A multimethodological approach using conventional techniques such as bacterial isolation and physiological studies, together with molecular genetics, will be necessary to fully develop the study of microbial ecology. The bacterial community can be studied using several approaches: first, a structural approach, attempting to study the entire soil bacterial community; second, the relationships between populations and the processes that regulate the system; and finally, a functional approach.

Recent research has shown that, within a bacterial population, cells are not isolated from each other but communicate to coordinate certain activities. This communication is key to their survival since microbial success depends on the ability to perceive and respond rapidly to changes in the environment. Bacteria have developed complex communication mechanisms to control the expression of certain functions in a cell density-dependent manner, a phenomenon termed as quorum sensing (QS).

Quorum sensing confers an enormous competitive advantage on bacteria, improving their chances to survive as they can explore more complex niches. This mechanism is also involved in the infection ability of some plant bacterial pathogens. Bacterial communication by quorum sensing is based on the production and release of signal molecules into the medium, termed autoinducers, concentration being proportional to cell density. When bacteria detect the signal molecule at a given concentration, the transcription of certain genes regulated by this mechanism is induced or repressed.

There are many microbial processes regulated by quorum sensing, including DNA transference by conjugation, siderophore production, bioluminescence, biofilm formation and the ability of some bacteria to move, called swarming. Recent studies have shown the importance of this type of regulation mechanism in putative beneficial bacterial traits for the plant, such as plant growth promotion, protection against pathogens or saline stress protection.

In addition, coevolution studies of plants and bacteria have determined that some plants release molecules, which mimic acyl homoserine lactones (AHLs) and even enzymes that are able to degrade the AHL molecule in root exudates. Somehow, plants have learned the language of bacteria and use it for their own benefit. Some studies have discovered that this behaviour leads to defense against plant bacterial pathogens, altering or blocking communication among bacteria, thus dramatically reducing their infection efficiency.

Efficiency of PGPR

Bacteria inhabiting the rhizosphere and beneficial to plants are termed Plant Growth Promoting Rhizobacteria (PGPR). Thus, the rhizosphere of wild plant species appears to be the best source from which to isolate plant growth promoting rhizobacteria. A putative PGPR qualifies as PGPR when it is able to produce a positive effect on the plant upon inoculation, hence demonstrating good competitive skills over the existing rhizosphere communities. Generally, about 2–5% of rhizosphere bacteria are PGPR.

Some PGPR have been produced commercially as inoculants for agriculture, but it must be borne in mind that the inoculation of these bacteria in soil may affect the composition and structure of microbial communities, and these changes must be studied since they have, at times, been related to the inefficiency of biofertilisers when applied to plant roots. On the contrary, many studies have tested the efficiency of PGPR in various conditions, observing that PGPR are efficient under determined conditions only.

Knowledge of the structure of rhizosphere microbial communities and their diversity, as related to other essential processes within the system such as complexity, natural selection, interpopulational relations, succession or the effect of disturbances, is the key to a better understanding of the system and for the correct utilisation of PGPR in biotechnology. Taking all of the above into consideration, it appears that quorum sensing can be a very useful tool in agriculture, with the potential to prevent bacterial pathogen attack and improve PGPR performance. There already exist transgenic plants that have been engineered to produce high levels of AHLs or an enzyme capable of degrading AHLs and that have demonstrated considerable capacity in blocking pathogen infection or altering PGPR performance.

Rhizosphere Microbial Structure

The bacterial community can be studied through two approaches: structural and functional. To understand the structural approach, we must know the groups of individuals, their species and abundance. Traditionally, this has been done by extracting microorganisms from the system, culturing them in the laboratory and performing many morphological, biochemical and genetic tests. Bacteria extraction methods require a dispersing agent to disintegrate the links among cells and need to be performed using either physical or chemical agents or a combination of both.

When handling bulk soil, rhizosphere soil and plant roots, dispersion methods need to be used owing to the intimate relationship between bacteria and the substrate. The efficiency of these methods is evaluated by comparing the microbial biomass of the original substrate before and after extraction. However, microbial biomass is difficult to calculate. There are several ways to approach these parameters including direct counting under a microscope, microbial respiration, ATP level assay, counting viable cells with the most probable number (MPN), using biomarkers such as lipids and soil fumigation with chloroform.

After extracting bacteria, several simple methods can be applied to isolate and count soil bacteria, such as growing them in a nonselective medium to obtain the total viable count (TVC). The data obtained with this method are expressed as colony forming units (CFUs). These studies, in which bacteria are grown on plates, are used to calculate the soil bacterial diversity, by observing the number and abundance of each species. Diversity indexes, such as the Shannon index (H), the Simpson index and the equitability index (J), have all been used to describe the structure of communities from a mathematical viewpoint.

The percentage of culturable microorganisms in soil is very low; however, some researchers estimate this at only 10%, while others suggest 1% or even lower. Because of the limitation of some methods, techniques in which it is not necessary to culture microorganisms on plates are required. One such technique is the phospholipid fatty acid analysis (PLFA). Phospholipids are integrated in the bacterial cell membranes. Different groups of microorganisms possess different fatty acid patterns. It is not usually possible to detect specific strains or species, but changes in the concentration of specific fatty acids can be correlated to changes in specific groups of microorganisms.

Another approach to nonculturable diversity is through techniques of molecular genetics, which, in the past 20 years, has revealed new information about soil microbial communities. Techniques include DNA and/or RNA hybridisation, polymerase chain reaction (PCR), ribosomal RNA sequencing, G+C percentages and DNA reassociation between bacteria in the community. At present, the most notable techniques are temperature gradient gel electrophoresis (TGGE) and denaturing gradient gel electrophoresis (DGGE), both based on the direct extraction of DNA or RNA from soil; the amplification of this DNA (by means of PCR), followed by electrophoretic separation in a temperature gradient for the former or by using chemical denaturing substances for the latter.

These techniques allow the separation of DNA fragments of exactly the same length but with different sequences, based on their melting properties. Other techniques include restriction fragment length polymorphism (RFLP), techniques related to the analysis and cutting of different restriction enzymes (amplified ribosomal DNA restriction analysis, ARDRA) or cloning the rDNA 16S and then sequencing. The use of microarrays is also an emerging technique with a promising future, which permits the identification of specific genes. Each of the methods described above possesses its own distinctive advantages and disadvantages. Generally, the more selective the method, the less able it is to detect global changes in communities and vice versa. Using these tools can provide an estimate of the microbial diversity in the soil.

Ecology and biodiversity of PGPR

In the last few years, the number of PGPR that have been

identified has seen a great increase, mainly because the role of the rhizosphere as an ecosystem has gained importance in the functioning of the biosphere and also because mechanisms of action of PGPR have been deeply studied. Currently, there are many bacterial genera that include PGPR among them, revealing a high diversity in this group. A discussion of some of the most abundant genera of PGPR follows to describe the genetic diversity and ecology of PGPR.

Free nitrogen-fixing bacteria were probably the first rhizobacteria used to promote plant growth. Azospirillum strains have been isolated and used ever since the 1970s when it was first used. This genus has been studied widely, the study by Bashan et al. being the most recent one reporting the latest advances in physiology, molecular characteristics and agricultural applications of this genus. Other bacterial genera capable of nitrogen fixation that is probably responsible for growth promotion effect, are Azoarcus sp., Burkholderia sp., Gluconacetobacter diazotrophicus, Herbaspirillum sp., Azotobacter sp and Paenibacillus (Bacillus) polymyxa.

These strains have been isolated from a number of plant species such as rice, sugarcane, corn, sorghum, other cereals, pineapple and coffee bean. Azoarcus has recently gained attention due to its great genetic and metabolic diversity. It has been split into three different genera (Azovibrio, Azospira and Azonexus). The most distinctive characteristic of these genera, which particularly differentiates them from other species, is their ability to grow in carboxylic acids or ethanol instead of sugars, with their optimum growth temperature ranging between 37 and 42 °C. Azoarcus is an endophyte of rice and is currently considered the model of nitrogen fixing endophytes.

Ninety-five percent of Gram-positive soil bacilli belong to the genus Bacillus. The remaining 5% are confirmed to be Arthrobacter and Frankia. Members of Bacillus species are able to form endospores and hence surviveunder adverse conditions; some species are diazotrophs such as Bacillus subtilis, whereas others have different PGPR capacities, as many reports on their growth promoting activity reveal.

Among Gram-negative soil bacteria, Pseudomonas is the most abundant genus in the rhizosphere, and the PGPR activity of some of these strains has been known for many years, resulting in a broad knowledge of the mechanisms involved. The ecological diversity of this genus is enormous, since individual species have been isolated from a number of plant species in different soils throughout the world. Pseudomonas strains show high versatility in their metabolic capacity. Antibiotics, siderophores or hydrogen cyanide are among the metabolites generally released by these strains. These metabolites strongly affect the environment, both because they inhibit growth of other deleterious microorganisms and because they increase nutrient availability for the plant.

Rhizosphere

Among the groups that inhabit the rhizosphere are rhizobia. Strains from this genus may behave as PGPR when they colonise roots from nonlegume plant species in a nonspecific relationship. It is well known that a number of individual species may release plant growth regulators, siderophores and hydrogen cyanide or may increase phosphate availability, thereby improving plant nutrition. An increase in rhizosphere populations has been reported after crop rotation with nonlegumes, with this abundance benefiting subsequent crops.

Microbial activity and functional diversity

The classical approach to determining functional diversity is to use culturable bacteria grown on a plate and subject them to selected biochemical tests. Another method involves analysing bacterial growth rate on a plate, which is considered as an indicator of the physiological state of the bacteria in the environment, the availability of nutrients and the adaptation strategy. It is known that culturable bacteria are scarce in soil but are considered responsible for the most important chemical and biochemical processes.

This is based on the fact that nonculturable bacteria are mostly dwarfs., measuring less than 0.4 mm in diameter and are considered as dying forms with almost no activity. Bååth studied the incorporation of radioactive precursors of DNA ($[H^3]$-thymidine, to assess population growth), and proteins (L-$[C^{14}]$-leucine, to assess population activity) in various fractions of soil filtrates. His research revealed that the culturable bacteria fraction (the larger size) is responsible for most of the growth and activity of the soil communities, whereas the fraction of cells less than 0.4 mm, considered nonculturable, had little importance in the metabolism and soil activity.

Finally, using the PLFA technique, it has been demonstrated that there are no significant differences between the phospholipid fatty acids of bacteria in soil and bacteria culturable from this soil. In contrast, other authors state that in rhizospheric communities, there are some difficulties in culturing groups of bacteria present in low densities that are metabolically very active; they can synthesize high amounts of proteins, use different substrates and are believed to be important in fundamental processes in the soil. These bacteria are called keystone species, some of which include Nitrosomonas and Nitrobacter, playing a very important role in the nitrogen cycle.

At present, enzymatic activity measurement is one of the more widely used techniques to determine microbial diversity, in which it is possible to perform studies with a specific enzyme. An other approach is to use Biolog plates, which permit microbial communities to be characterised according to their physiological profile (community-level physiological profile, CLPP) calculated from the different útilisation patterns of many carbon and nitrogen sources, determined by a redox reaction that changes color after inoculation and incubation of the microbial communities.

New approaches such as the search for new catabolic, biosynthetic or antibiotic functions in soil samples are required to identify new, potentially nonculturable genotypes. The cloning and sequencing of large DNA fragments (BAC library) will provide researchers with information about the metabolic diversity of nonculturable and culturable strains in the future and also provide important information on ecological laws and the operation of the soil ecosystem.

Undoubtedly, future studies on soil communities will involve microarray techniques that will permit the study of differences in the structure of communities, identifying groups that are active or inactive during a specific treatment leading to the identification of strains isolated from different environments and explaining differences or similarities in the operation of niches. These techniques are complemented with transcriptomic techniques, based on the description of the activity of a gene by its expressed mRNA, and the proteomic approximation.

Activity and effect of PGPR

Some researchers approach the study of biochemical diversity in soil by identifying biochemical activities related to putative physiological PGPR traits in bacteria isolated from the rhizosphere (Table 1).

Microbial activity in the rhizosphere indicates how metabolically active the microbial communities are. Using PGPR as inoculants in soil, besides altering the structure of the communities, will also influence microbial activity, and this could be related to the survival of the PGPR in the environment. Some of the factors influencing the survival and activity of bacteria in the rhizosphere are physical (texture, temperature and humidity), while others are chemical, such as pH, nutrient availability, organic matter content and, above all, interactions with other rhizosphere microorganisms.

Table 1: Frequency of physiological PGPR traits in the mycorrhizosphere of P. pinaster and P. pinea and the associated mycosphere of L. deliciosus.

	P. pinaster		*P. pinea*	
PGPR trait	*Mycorrhizo-sphere*	*Mycosphere*	*Mycorrhizo-sphere*	*Mycosphere*
Aux (%)	14	0	50	42
Aux+PDYA (%)	0	0	0	2
Aux+CAS (%)	0	3	11	2
Aux+ACC (%)	0	0	7	0
Aux+CAS+PDYA (%)	0	3	0	0
PDYA (%)	47	35	11	32
PDYA+ACC (%)	3	0	0	0
CAS (%)	36	40	14	11
CAS+PDYA (%)	0	3	0	0
CAS+PDYA+ACC (%)	0	3	0	0
ACC (%)	0	13	7	11

The interaction with the biotic factor is very important because PGPR must occupy a new niche, adhering to the plant roots, and the inoculum must compete for available nutrients released, essentially, by the root exudates, maintaining a minimum population able to exert its biological effect. Studies of characterisation of the soil microbial community activity are conducted using various techniques, such as thymidine ($[H^3]$) incorporation, radioactive DNA precursors to assess population growth and leucine (L-$[C^{14}]$) radioactive protein precursor to assess the metabolic activity of the population. Stable isotope probing (SIP), based on radioactive labelling of different substrates, is considered to have enormous potential. A further approach to quantifying the activity in the rhizosphere is by means of SIR.

Screening Strategies

The rhizosphere of wild populations of plants is proposed as one of the optimal sources in which to isolate PGPR. This is because of the high selective pressure a plant exerts in this zone. The plant selects, among others, beneficial bacteria. In the screening of PGPR, the different soil types, plant species, seasons and the plants physiological moment must be considered to ensure the successful isolation of putative beneficial rhizobacteria. The first step in obtaining a PGPR is the isolation of rhizospheric bacteria. It is generally accepted that the rhizosphere is the soil volume close to the roots.

To collect this soil fraction, the root is normally shaken vigorously and soil still adhering is collected as the rhizosphere. Depending on the type of study, the root containing the endophyte bacteria is included, as some have been described as PGPR. Other researchers refer to the rhizosphere as the soil adhering to the roots after they have been washed under running water. Rhizobacteria extraction starts with the suspension of soil in water, phosphate buffer or saline solution. Some compounds such as pyrophosphate are effective for soil disgregation, but can alter cell membranes.

Sample dispersion is made with chemical dispersants such as chelants that exchange monovalent ions (Na^+) for polyvalent cations (Ca^{2+}) of clay particles, reducing the electrostatic attraction between the soil and the bacterial cells. Various researchers have used ionic exchange resins derived from iminodiacetic acid, for example, Dowex A1 or Chelex-100. Other dispersants are Tris buffer or sodium hexametaphosphate. Detergents are used because the microbial cells present in the treated sample adhere by extracellular polymers to the soil particles.

MacDonald demonstrated that using detergents together with Dowex A1 increased the microbial extraction from soil to 84%. This method was modified later by Herron and Wellington, replacing Dowex with Chelex-100 and combining with polyethylene glycol to dissolve and separate the phases. Other chemical solvents used in extraction protocols are Calgon at 0.2% for the extraction of bacteria from soil in studies of bacterial counts with acridine orange, citrate buffer used in studies of membrane phospholipids from soil microbes and Winogradsky solution for microbial diversity studies using molecular techniques (ARDRA, DGGE or REP-PCR) or phenotypical tests (Biolog).

Chemical extraction methods may be combined with physical methods, and these can be divided into three categories: shaking, mixing and ultrasonics. Shaking is probably the least efficient method but adequate for sensitive bacteria or bacteriophages. Techniques based on homogenisation could damage some groups of bacteria, such as Gram-negative bacteria, and extraction would be selective.

A combined method of grinding and chemical dispersants would be more effective. Ultrasonic treatments are the best among methods used to break the physical forces between soil particles. In clay soils, pretreatment of the sample is necessary; however, most sensitive bacteria, such as Gram-negative ones, could be damaged. This effect can be avoided using less aggressive ultrasonic treatments.

After rhizobacterial isolation, a screening of the putative PGPR is performed using two different strategies:

(a) Isolation, to select putative bacteria beneficial to the plant using specific culture media and specific isolation methods. For example, Founoune et al. isolated Pseudomonas fluorescens from the Acacia rhizosphere as a species described as PGPR.

(b) After isolation of the maximum number of bacteria to avoid the loss of bacterial variability, different tests are performed to reduce the various types of bacteria chosen, so that only the putative beneficial ones remain. The test is performed in vitro to check biochemical activities that correspond with potential PGPR traits. Genetic tests may also be performed to remove genetic redundancy, that is, select different genomes that may have different putative beneficial activities.

Among the biochemical tests used to find putative PGPR traits, the most common are the following:

(i) test for plant growth regulator production;
(ii) the ACC (1-aminocyclopropanecarboxylic acid) deaminase test; this enzyme degrades the ethylene precursor ACC, causing a substantial alteration in ethylene levels in the plant, improving root system growth;
(iii) phosphate solubilisation test, phosphate solubilisation may improve phosphorous availability to the plant;
(iv) siderophore production test, which may improve plants iron uptake;
(v) test for nitrogen-fixing bacteria to improve the plants nitrogen nutrition; and
(vi) test for bacteria capable of producing enzymes that can degrade pathogenic fungi cell walls preventing plant diseases.

The most common genetic techniques are PCR-RAPD (randomly amplified polymorphic DNA, ERIC-PCR, BOX-PCR and REP-PCR. They all compare bacterial genomes and establish a homology index among them. These techniques allow the formation of groups of bacteria with very similar genomes and thus with supposed similar PGPR abilities. This approach of testing in vitro abilities has been proved to be an effective strategy to isolate PGPR;

however, there are limitations. Some of the biochemical traits shown in vitro are inducible; that is, they are expressed in certain conditions but not in others.

Therefore, a bacterial PGPR trait could be expressed in the laboratory in a culture media but not in the rhizosphere. This is true of PGPR traits related to plant nutrition, such as phosphate solubilisation and siderophore production, that are not expressed in phosphorous-rich and iron-rich soils, respectively. There are also problems with a bacterial property called phase variation, which produces strong genetic variations in bacteria by an enzyme called site-specific invertase. Hence, when these genetic variations occur, a strong phenotypical change occurs. It may be the case that a bacterial culture exhibits a PGPR trait, but after a time does not because of these phase variations.

After the screening process, the PGPR potential shown in vitro should be tested to ensure the same effect occurs in the plant. Root colonisation is a necessary requirement for the bacteria to exert its effect. PGPR inoculation in distinct plant species sometimes produces erratic results; however, the factors leading to failure are unclear. The competitive interactions in the rhizosphere are not well known. Independent of the factors that lead to good colonisation, an inoculum screening is required to assess its impact on the rhizosphere.

The introduction of a putative PGPR may alter the microbial rhizosphere communities, and this is indirectly related to plant fitness. The introduced population can establish itself in the rhizosphere without changing the microbial communities or it may not establish itself but change the communities. It is necessary to know how the inoculum is going to evolve to calibrate the potential risks of introducing these microorganisms, whether genetically modified or not. Several researchers have reported the alteration of these communities as a key to PGPR efficiency.

Biological trials may be performed in a sterile system to assess bacterial root colonisation abilities. In addition, they may also be performed in either a nonsterile system or in a natural system (field trials). Competitiveness of the putative PGPR strain is a necessary requirement for colonisation and to demonstrate the biological effect. On the contrary, the alteration of microbial communities present in a natural rhizosphere can also be studied in these types of systems. A range of screening processes appeared in the recent literature. Cattelan et al. tested several biochemical activity indicators for putative PGPR abilities in 116 bacterial strains selected from bulk soil and the rhizosphere of soybean.

The indicators tested were phosphate solubilisation, indole acetic acid production, siderophore production, chitinase, β-1,3-glucanase, ACC deaminase and cyanide production, putative free-living nitrogen-fixing bacteria and fungi growth inhibition. Twenty-four strains showed one or more of these activities and were assayed for traits associated with biocontrol, inhibition of rhizobial symbiosis and rhizosphere competence. These were finally tested for promotion of soybean growth. Six of the eight isolates tested positive for 1-aminocyclopropane-1-carboxylate deaminase production, four of the seven isolates were positive for siderophore production, three of the four isolates tested positive for β-1,3-glucanase production and two of the five isolates tested positive for phosphate solubilisation, increasing at least one aspect of early soybean growth.

More examples of screening processes are those performed in the laboratory of Dr Gutiérrez Mañero, where PGPR have been isolated from the rhizosphere of wild plant species. For example, a screening for PGPR was performed in the mycorrhizosphere of wild populations of Pinus pinea and P. pinaster and in the mycosphere of associated Lactarius deliciosus, being the targeted microorganisms that are able to enhance establishment of mycorrhisation. Of the 720 isolates, 50% were tested for ACC degradation, auxin and siderophore production and phosphate solubilisation.

One hundred and thirty-six isolates showed at least one of the evaluated activities. After PCR-RAPD analysis, 10 groups were formed with 85% similarity when all isolates were considered. One strain of each group was tested to see if it improved pine growth and eight were found to be effective. PGPR have also been isolated from the rhizosphere of Nicotiana glauca to improve the performance of Lycopersicon esculentum, a plant from the same family. The rationale was that the rhizosphere of wild populations of N. glauca would be a good source for putative PGPR able to induce systemic resistance and hence to be used in reducing chemical inputs of pesticides.

A screening of 960 strains in the rhizosphere of Nicotiana, grown in three different soils (calcareous, quaternary and volcanic), was performed in both hot and cold seasons to isolate PGPR associated with this genus. A subset of 442 isolates composed of the most abundant parataxonomic groups was characterised based on their metabolic activities regarded as putative PGPR traits related to defense (siderophore and chitinase production). Fifty percent tested positive for both traits and were tested for growth promotion of *L. esculentum* seedlings and induction of resistance against Fusarium and Xanthomonas. The results were positive for 30 strains in growth, while only 6 enhanced resistance against foliar pathogen.

Other researchers have also isolated a large number of PGPR bacteria. From the rhizosphere soil of wheat plants grown at different sites, 30 isolates that showed prolific growth on agar medium were selected and evaluated for their potential to produce auxins in vitro. A series of laboratory experiments conducted in two cultivars of wheat under gnotobiotic (axenic) conditions exhibited increases in root elongation (up to 17.3%), root dry weight (up to 13.5%), shoot elongation (up to 37.7%) and shoot dry weight (up to 36.3%) of inoculated wheat seedlings.

A positive linear correlation between in vitro auxin production and increase in growth parameters of inoculated seeds was found. Furthermore, auxin biosynthesis in sterilised versus nonsterilised soil inoculated with four selected PGPR was also monitored and demonstrated the superiority of the selected PGPR over indigenous microflora. Field experiments showed an increase of up to 27.5% over the control using these PGPR. Finally, researchers such as Eleftherios et al. have performed a direct screening to obtain endophytic bacteria able to protect against the pathogenic fungi *Verticillium dahliae*.

Four hundred and thirty-eight bacteria were isolated from tomato root tips and 53 of these were found to be antagonistic against *V. dahliae* and several other soilborne pathogens in dual cultures. Significant biocontrol activity against *V. dahliae* in glasshouse trials was demonstrated in 3 of 18 evaluated antagonistic isolates. Finally, two of the most effective bacterial isolates, designated as K-165 and 5-127, were tested for rhizosphere colonisation ability and chitinolytic activity, with both giving positive results.

REFERENCES

Belimov, A. A., Kojemiakov, A. P. & Chuvarliyeva, C. V., "Interaction between barley and mixed cultures of nitrogen fixing and phosphate-solubilising bacteria," *Plant and Soil*, 173, 29-37, 1995a.

Khaosaad, T., Garcia-Garrido, J. M., Steinkellner, S. & Vierheilig, H., "Take-all disease is systemically reduced in roots of mycorrhizal barley plants," *Soil Biology and Biochemistry*, 39, 727-734, 2007.

Singh, S. & Kapoor, K. K., "Inoculation with phosphate-solubilising microorganisms and a vesicular-arbuscular mycorrhizal fungus improves dry matter yield and nutrient uptake by wheat grown in sandy soil," *Biology and Fertility of Soils*, 28, 139-144, 1999.

11

Biotechnology and Biofertilisation

All human beings depend on agriculture that is required to produces food of the appropriate quality and in required quantities. Domestication of plants and animals found in the wild, combined with gradual long-term changes in their qualitative traits, were the first attributes of agriculture. Domestication, followed by food storage, coincided with the growth of microorganisms. The "green revolution", increased wheat production 10-fold in India, Pakistan and several countries in south East Asia, thereby feeding three times as many people. However this revolution has already been exploited to its limits, and alternative solutions are required to be embarked upon through integration of classical breeding and advanced techniques in crop production and biotechnology to create "evergreen revolution" on sustainable basis.

CURRENT FARMING PRACTICES

Current farming practices heavily rely on the use of chemical fertilisers with major emphasis on N and P fertilisers while K and micronutrients are generally not applied. Besides, almost all the crop residues are removed. This, coupled with planting of high yielding varieties, has led to decline in soil fertility and organic matter (< 0.5%) and crop production potential of soils. All the agricultural soils in Pakistan are deficient in nitrogen, 80-90% in phosphorus, and 30-40% in potassium. Among micronutrients, field scale deficiencies of economic significance prevail in zinc, boron, and iron.

According to a recent survey, 94% farmers used NPK fertilisers whereas 32.6% farmers use farmyard manure, 1.6% micronutrients, 2.7% crop residues and less than 1% use bio-fertilisers. On the whole, the fertiliser use in Pakistan is increasing day by day but soil fertility and organic matter levels are decreasing with the passage of time. Total consumption of nutrients has reached 3.8 million tons of which nitrogen accounts for 77% and phosphorus 22%. This imbalanced nutrition is not only impacting crop productivity but it is leading to serious problem of soil degradation and environmental concerns as well.

During the past 20 years, population has registered a growth of 2.4% per year, and fertiliser use has increased at 4.7% year. On the other hand, cropped area availability per capita has decreased from 0.21 to 0.14 hectares, and yields of major crops have increased by only 1.5-1.7% per year. During the same period, fertiliser efficiency has decreased, for wheat for example, from over 20 to less than 12 kg of grain per kg of fertiliser used. Thus increase in fertiliser consumption is not fully translated in crop production.

Crop yields continue to remain low which can be increased at least two-fold by adopting proper soil and water management practices, balanced crop nutrition involving the use of all possible sources of nutrients, improved crop production technology etc. Under the situation, biofertilisers offer great potential for not only increasing improving soil fertility and organic matter levels but also provide for efficient use of various inputs (water and nutrients) and for increasing crop production on sustainable basis.

Biofertilisers

Biofertiliser is a material containing microorganism(s) added to a soil to directly or indirectly make certain essential elements available to plants for their nutrition. Various sources of biofertilisers include nitrogen fixers, phytostimulators, phosphate solubilising bacteria, plant growth promoting rhizobacteria etc. Considerable progress has been made over the past two decades in evaluation of these technologies and development of application methods, yet we stand far away from adoption of these technologies.

The simplest way of harnessing the benefits include planting of legumes which can fix atmospheric nitrogen to the extent of 250- 300 kg/ha/yr, besides improving soil fertility and added benefit to succeeding crops. Inclusion of legumes and green manure crops in the cropping systems can, therefore, substantially reduce nitrogen fertiliser requirements, increase fertiliser use efficiency and reduce production costs while improving soil quality and organic matter contents in an environment friendly manner.

The discussion about quality and soil organic matter would be incomplete without discussing the positive contribution of organic wastes of crop, animal and industrial origin. Complete crop residue removal for fodder and fuel happens to be a norm in Pakistan, and application of animal manure and other organic wastes is also quite limited. Thus, depletion of soil organic matter stock from the root zone has adversely affected the soil productivity and environmental quality. In simple terms, poor farmers have passed on their suffering to the land through extractive processes. They cultivate marginal soils with marginal inputs, produce marginal yields and perpetuate marginal living and poverty.

The situation deserves concerted efforts and appropriate strategies and policy decisions to overcome the negative impact on crop production and food security environment. Of the available organic sources, farm yard manure is a potential source but almost 50% of it is not collected, and approximately half of that is used as fuel. So in essence only 25% of farm manure finds its way to the agriculture fields. Based on different assumptions, it is estimated that 1.5 million tons of nutrients are available from FYM, in addition to 0.185 million tons from poultry manure.

The Pakistan sugar industry produces 1.2 million tons of filter cake every year which is a rich source of organic matter, and major and micronutrients. Application of these organic wastes and including biofertilisers in the farming systems can help maintain consistently high yields through improvements in water and nutrient use efficiencies and soil biotic activity and soil organic matter levels. An increase of 1 ton of soil organic carbon pool of degraded soils may increase crop yield by 20-40 kg/ha for wheat and similar increases for other crops, besides enhancing food security.

Organic Agriculture

"Organic agriculture is a holistic production management system which promotes and enhances agroecosystem health, including biodiversity, biological cycles, and soil biological activity. It emphasises the use of management practices in preference to the use of off-farm inputs, taking into account that regional conditions and locally adapted systems. This is accomplished by using, where possible, agronomic, biological, and mechanical methods, as opposed to using synthetic materials, to fulfil any specific function within the system".

Organic agriculture is a system that relies on ecosystem management rather than external agricultural inputs. It is a system that begins to consider potential environmental and social impacts by eliminating the use of synthetic inputs, such as synthetic fertilisers and pesticides, veterinary drugs, genetically modified seeds and breeds, preservatives, additives and irradiation. These are replaced with site-specific management practices that maintain and increase longterm soil fertility and prevent pest and diseases.

There is increasing awareness among the farmers regarding the benefits of organisms in crop production and in improving soil health and nutrition status. Some farmers have already started converting their farms into organic agriculture where quality of the product and the environmental concerns stand a priority. Production levels are often low under organic agriculture compared to conventional system based on agrochemicals but the farm produce from organic agriculture fetches better price as export commodity.

Plant Biotechnology

During the last two decades, the new biotechnologies have been adopted to agricultural practices which have opened new vistas for plant utilisation for enhancement in agriculture productivity in a sustainable manner. This will continue and intensify in the next decade. Plant biotechnology – especially in-vitro regeneration and cell biology, DNA manipulation and genetic modification of biochemical pathways—is changing the plant scene in three major areas i.e., growth and development control, protecting plants against the ever-increasing threats of abiotic and biotic stress, and expanding the horizons by producing specialty foods, biochemicals and even bio-pharmaceuticals.

Achievements today in the biotechnology have already surprised all previous expectations, and the future is even more promising. The full realisation of the agricultural biotechnology revolution depend on both continued successful and innovative research and development activities. Biotechnology should be fully integrated as an aid to classical breeding techniques, through bioengineering and molecular biology interventions techniques and by introduction and investment of engineering microorganisms into agricultural production systems. It must be realised that only about 12% of the world's land surface is used to grow crops, and the agricultural areas required to support food production is going to drop to from – 0.44 ha / capita in 1961 to 0.15 ha / capita, in view of the population growth.

The challenge for the agricultural sector during the next few decades is therefore clear, double the food production by 2025, on less per capita land, with less water, under increasingly challenging environmental conditions. The situation is further complicated by the fact that in spite of the heavy use of agrochemicals, modern agriculture still loses nearly 42% of crop productivity to competition with weeds and to pests and pathogens, and an additional 10-30% to post-harvest losses due to a variety of factors, especially in the developing countries where storage conditions are poor.

During the 20th century traditional plant breeding has brought about enormous increases in crop productivity. However, plant improvement by hybridisation is slow and is restricted to a very small gene pool owing to natural barriers to crossability. Beginning in the early 1980s, advances in plant cell culture and genetic transformation have overcome these barriers by making it possible to transfer defined genes into all major food crops, including cereals, legumes, cassava, potato, and many vegetable fruits. The entire global gene pool whether it be plant, animal, bacterial or viral- is available for utilisation.

The first genes that have been integrated into crop species provide resistance to non-selective and environment friendly herbicides, and many pests and pathogens. Increasingly large acreages of transgenic maize, soybean, potato, tomato and cotton are being commercially grown for human use and consumption. Carefully planned introduction of such crops on a worldwide scale would greatly help in reducing or even elimination of the enormous crop losses attributed to weeds, pests and pathogens. The use of such crops will also have a beneficial effect on the environment by significantly reducing the use of agrochemicals. Other genes for improving crop productivity, and manipulating starch / protein / oil quality and quantity, resistance to environmental stresses such as temperature and drought are also being isolated and studied to produce second generation transgenic crops.

The biotechnology is required to be geared up and directed to develop biomaterials aimed at enhanced productivity to meet the food requirements of galloping population on sustainable basis in years to come. Two major categories of biomaterials could be direct improvement and modification of specialised constituents of plant origin, and manufacture in plants of non-plant compounds. Biotechniques, mostly based on the engineering of metabolic pathways, are now available to modify many plant constituents that are used in the food, chemical and energy industries to convert them into their bioforms.

This includes many "primary" metabolites: carbohydrates (starch synthesis, yield and allocation, production of high-amylose or high-amylopectin starch, increased sucrose synthesis for sugar industry, enhanced production etc.), fructan production, proteins (improvements of amino acid composition and protein content), oils and fats (ratio of saturated to nonsaturated fatty acids, increased content of specific valuable fatty acids like uric acid, ricinoleic acid and others). It is now generally realised that the enhanced agriculture productivity could only be attained through the practice of better plant breeding which will continue to move forward as progress is being made in advanced techniques through biotechnology endevaours and its consequent limitations associated with use of synthetic chemical fertilisers as well as insecticides and pesticides.

Scientists in the future will not only be forced to hone their traditional skills but will also have to integrate them with knowledge and experience to coincide with recent advancement and technologies derived from cellular and molecular concept and approaches. For worldwide crops such as potato, rice, maize and in the future wheat, as well as for high value vegetable crops, the future of agriculture and scientists is associated with the use of combination of practices involving traditional breeding, genetic engineering and tissue culture techniques. In addition to increasing reliance on breeding, agriculture will also depend on biocontrol to complement and make possible the use of chemicals that are compatible with intelligent management of the natural resources essential for a sustainable yet highly productive agriculture.

Plant tissue culture technique an aspect of agriculture, not only provides essential way for the clonal propagation of many agriculturely important crops such as woody ornamentals and vegetable plants but they are also the basis for the production of transgenic plants and are at the forefront of the recent studies in biotechnology and plant physiology, developmental biology and biochemistry. Spectacular progress in our knowledge of plant sciences, relevant to agriculture and breeding, can be expected from the integration of more traditional sciences with molecular and cellular concept and techniques.

Sustainability, however, would never be achieved as long as our agriculture practices continue beyond the carrying capacity of the ecosystem through indiscriminate use and abuse of agricultural chemicals. The rapid growth of agriculture productivity in chemical farming systems in the wake of intensive farming and population growth needs a serious rethinking due to continued soil degradation, loss of fertility and environmental pollution. Environmental concerns from agricultural activities jeopardise agriculture growth in several countries of Asia, Africa and subsaharan region.

Problems associated with the agriculture production in the developed world and underproduction in developing countries therefore, necessitates an assessment of the present status of agriculture. It is time to develop new farming systems committed to environment friendly and sustainable approaches, producing healthy food free of agrochemical residues. Ecologically oriented farming routines including biofertilisation techniques need to be developed and embarked upon within the framework of recent achievements in environmental biotechnology, the most important of which is the clean farming system which is increasingly acknowledged as a potential solution to copious problems of present world agriculture. It is a farming system, which aims at evading the routine use of agricultural chemicals and reducing their rates of application.

Clean farming system is associated with four environmental biotechnologies, i.e., recycling of composted organic waste, fortifying the rhizosphere soil with biofertilisers, encouraging the use of biopesticides in agricultural practices and bioremediation of polluted agro-ecosystems. Increasing realisation of the ill effects of long sustained, exclusive use of chemical fertilisers, and consistent growing demand from the consumers for quality, coupled with unsustainable productivity of crops, fruits and vegetables have fostered experimentation with some alternative cultural practices. Organic culture is claimed to be the most benign alternative.

Use of organic materials such as farmyard manure, cakes of plant origin, vermicompost, and microbial bio-fertilisers on one hand, and exploiting the synergism between fruit-vesicular arbuscular mycorrhizal fungus on the other hand, are important components of the bio-organic concept of fruit / crop cultivation. Mycorrhizae were observed to be highly effective in low fertility, coarse texture soils. Mycorrhizal-treated trees had better plant growth and uptake of nutrients like P, Ca, Zn, Cu, and Fe compared to non-mycorrhizal trees. Inoculation of soil with mycorrhizae also helps in regulating the water relations and carbohydrate metabolism of fruit trees / vegetables.

Phosphorus nutrition of mycorrhizae also helps in regulating the water relations and carbohydrates metabolism of crops. Phosphorus nutrition of mycorrhizal-treated fruit tree was best improved by using biofertilisation techniques and by adopting alternative farming systems through organic agriculture practices. Over the years the PGPR (plant growth promoting rhizobacteria) have gained worldwide importance and acceptance for agricultural benefits. These microorganisms are the potential tools for sustainable agriculture and the trend for the future.

Scientific researchers involve multidisciplinary approaches to understand adaptation of PGPR to the rhizosphere, mechanisms of root colonization, effects of plant physiology and growth, biofertilisation, induced systemic resistance, biocontrol of plant pathogens, production of determinants etc. Biodiversity of PGPR and mechanisms of action for the different groups: diazotrophs, bacilli, pseudomonads, and rhizobia are shown. Effects of physical, chemical and biological factors on root colonisation and the proteomics perspective on biocontrol and plant defence have also shown positive results.

Visualisation of interactions of pathogens and biocontrol agents on plant roots using autofluorescent protein makers has provided more understanding of biocontrol processes with overall positive consequences. This is a very interesting topic which is part of agricultural microbiology. It is all about biofertilisers. Life is much more than chemicals like K,P,N. Even more than that, Pure Chemicals kills the life inside our mother Earth! There are millions of Microscopic Organisms near the plants that conform a in a micro enviroment that provides nutrients to the plants also helps to keep the water and retain the nutrients in the soil easy available to the plants. *Bio Fertiliser* is a 100% natural and organic fertiliser that helps to provide and keep in the soil all the nutrients and microorganisms required for the benefits of the plants.

BIOFERTILISERS

A biofertiliser is a large population of a specific or a group of beneficial microorganisms, millions and billions of them incorporated aseptically into sterile carrier materials such as peat, lignite or charcoal. Such material is generally packed in plastic bags and sold to the farmers as biofertilisers for enhancing the productivity of soil either by fixing atmospheric nitrogen or by solubilising soil phosphorus or by stimulating plant growth through synthesis of growth promoting substances.

Biofertilisers are the most advanced biotechnology necessary to support developing organic agriculture, sustainable agriculture, green agriculture and non-pollution agriculture. This Bioorganic Fertiliser can increase the output, improve the quality and it is responsible for agriculture environment. Today, It has been widely used with excellent results in all kinds of plants and several countries. It is well known that the continued use and overuse of petrochemical based fertilisers and toxic pesticides have caused a detrimental effect to our soils, water supplies, foods, animals and even people.

The Biological Grower is more sensible with sustainable approach and employs the resources of both science and nature to allow better results in his production. For centuries, peat moss has been recognised soil bacteria, fungi, earthworms, and other bio-organisms to enrich the soil to produce safe, nutritious and abundant crops, an excellent fertiliser sources. We have developed all natural organic fertilisers made with pure peat moss. Biofertilisers contain a wide range of naturally chelated plant nutrients and trace elements, carbohydrates, amino acids and other growth promoting substances. Kelp acts as a soil conditioner by stimulating microbial activity in the soil which results in improved air-water relationships in soil, improved fertility and makes soil less prone to compaction and erosion.

Organic Growers who use kelp in their regular fertility program report increases in yield, quality, shelf life and resistance to environmental stresses such as drought, extreme heat, early frost, pest and disease problems. Organic fertilisers differ from chemicals fertilisers in that they feed your plants while adding organic material to the soil. Soils with lots of organic matter remain loose and airy, hold more moisture and nutrients, foster growth of soil organisms, and promote healthier plant root development. If only chemicals are added the soil gradually loses its organic matter and micro biotic activity.

As organic matter is used up, the soil structure deteriorates, becoming compact, lifeless and less able to hold water and nutrients. This results in increased amounts of chemical fertilisers needed to feed plants. We also like organic fertilisers because they're made from renewable resources; chemicals are not. The Bio Fertiliser is a premium natural fertiliser composed just with certified organic ingredients special for nutrient-poor Western soils. This organic fertilisers is unequaled in its ability to nourish the beneficial micro-organisms in the soil greatly increasing the soil's humus content and improving its ability to sustain and nurture healthy, more colorful plants.

Use by the handful when planting individual plants, broadcast and mix it deeply into the soil when planting flower beds or spread it around established plants and scratch it into the soil. It is also excellent for use in vegetable gardens, container plantings and as a compost-pile activator. One of the major concerns in today's world is the pollution and contamination of soil. The use of chemical fertilisers and pesticides has caused tremendous harm to the environment. Organic fertilisers differ from chemicals in that they feed your plants while adding organic material to the soil.

Soils with lots of organic matter remain loose and airy, hold more moisture and nutrients, foster growth of soil organisms, and promote healthier plant root development. If only chemicals are added the soil gradually loses its organic matter and microbiotic activity. As organic matter is used up, the soil structure deteriorates, becoming compact, lifeless and less able to hold water and nutrients. This results in increased amounts of chemical fertilisers needed to feed plants. We also like organic fertilisers because they're made from renewable resources; chemicals are not. Biofertilisers are organisms that enrich the nutrient quality of soil. The main sources of biofertilisers are bacteria, fungi, and cynobacteria.

The most striking relationship that these have with plants is symbiosis, in which the partners derive benefits from each other. Plants have a number of relationships with fungi, bacteria, and algae, the most common of which are with mycorrhiza, rhizobium, and cyanophyceae. These are known to deliver a number of benefits including plant nutrition, disease resistance, and tolerance to adverse soil and climatic conditions. These techniques have proved to be successful biofertilisers that form a health relationship with the roots.

Biofertilisers will help solve such problems as increased salinity of the soil and chemical run-offs from the agricultural fields. Thus, biofertilisers are important if we are to ensure a healthy future for the generations to come. The manufacturing process of biofertiliser includes the collection of bacterial strains, mother culture, multiplication, grinding and sterilisation of the carrier material, blending of the bacterial broth culture with carrier and packing. The various equipments/ instruments required for the manufacture of biofertilisers are—steam boiler, autoclaves, fermenters, blending machines and packing machines.

Laboratory equipments like laminar flow, BOD incubator, shakers, microscope, spectrophotometer, pH meter, refrigerator, sensitive balances, colony counter, hot air oven, deioniser, centrifuge and other routine laboratory glassware are also required. In view of the huge area under cultivation for pulses, oilseeds, cereals, millets, fibre, sugarcane and vegetable crops, there is enormous potential for biofertiliser marketing in India. About 460 million packets of inoculate of 200 g each are required for inoculating to various crops having a coverage of 184 million hectare in India. Looking into the future expansion of areas and evolution of better strains which would be required to be provided everywhere, the future expected demand of biofertiliser is to be in the order of 15,000 to 20,000 MT per year.

Mycorrhiza

Mycorrhizae are a group of fungi that include a number of types based on the different structures formed inside or outside the root. These are specific fungi that match with a number of favourable parameters of the host plant on which it grows. This includes soil type, the presence of particular chemicals in the soil types, and other conditions. These fungi grow on the roots of these plants. In fact, seedlings that have mycorrhizal fungi growing on their roots survive better after transplantation and grow faster.

The fungal symbiont gets shelter and food from the plant which, in turn, acquires an array of benefits such as better uptake of phosphorus, salinity and drought tolerance, maintenance of water balance, and overall increase in plant growth and development. While selecting fungi, the right fungi have to be matched with the plant. There are specific fungi for vegetables, fodder crops, flowers, trees, etc. Mycorrhizal fungi can increase the yield of a plot of land by 30%-40%. It can absorb phosphorus from the soil and pass it on to the plant. Mycorrhizal plants show higher tolerance to high soil temperatures, various soil- and root-borne pathogens, and heavy metal toxicity.

Legume and Rhizobium

Leguminous plants require high quantities of nitrogen compared to other plants. Nitrogen is an inert gas and its uptake is possible only in fixed form, which is facilitated by the rhizobium bacteria present in the nodules of the root system. The bacterium lives in the soil to form root nodules in plants such as beans, gram, groundnut, and soybean.

Blue - Green Algae

Blue-green algae are considered the simplest, living autotrophic plants, i.e. organisms capable of building up food materials from inorganic matter. They are microscopic. Blue-green algae are widely distributed in the aquatic environment. Some of them are responsible for water blooms in stagnant water. They adapt to extreme weather conditions and are found in snow and in hot springs, where the water is 85 °C.

Certain blue-green algae live intimately with other organisms in a symbiotic relationship. Some are associated with the fungi in form of lichens. The ability of blue-green algae tophotosynthesize food and fix atmospheric nitrogen accounts for their symbiotic associations and also for their presence in paddy fields. Blue-green algae are of immense economic value as they add organic matter to the soil and increase soil fertility. Barren alkaline lands in India have been reclaimed and made productive by inducing the proper growth of certain blue-green algae.

Azotobacter

Azotobacter is a free-living nitrogen-fixing bacterium, which is used as a biofertiliser in the cultivation of most crops. A specific strain of *Azotobacter chroococcum* has been isolated from the rhizosphere of mulberry plant and tested by the CSR&TI, Mysore under *in vivo* and *in vitro* conditions and has found to be very effective. The laboratory has recommended application of the *Azotobacter* biofertiliser at the rate of 20 kg per hectare per year and this assures a substantial saving of nitrogen fertiliser (about 50 % reduction in N requirement). This indirectly reduces the costs involved in the silkworm rearing aspects as well as the soil health is improved by increased microbial activity.

The process of mass multiplication of the microbial culture and packing in polythene bags along with a carrier (preferably lignite or coal) has been standardised. The process ensures a minimum of 108 viable cells of *Azotobacter* per gram of carrier on dry weight basis within 15 days of preparation and 107 viable cells before the expiry date (As per the BIS norms). The shelf life of the biofertiliser is six months from date of manufacture.

The Process requires locally available plant, machinery and raw materials for manufacture. The mass culture can be done by shake-flask or by using a fermenter depending on the investment capability. The culture is then mixed with the carrier material under hygienic conditions and packed immediately. The batch number and expiry date are immediately affixed on each bag. Normally, packets are available in 250 g, 500 g, 1.0 kg and 5.0 kg denominations.

Bio-fertilisers based on renewable energy source are cost effective supplement to chemical fertilisers, eco-friendly and can help to economise on the high investment needed for chemical fertiliser use as far as nitrogen and phosphorus are concerned.

Mycorrhizal Fungi

Mycorrhizal fungi are specialized microbes that form a close association with most plants that grow on land. Mycorrhizal fungi are the most researched group of beneficial fungi in science with over 15,000 publications on the subject. They naturally colonise roots and dramatically increase the root absorptive area resulting in an impressive increase in uptake of water and major and minor nutrients. Once plants become mycorrhizal, there is no need to reinoculate since the fungus becomes a permanent part of the root system. Plants with mycorrhizal roots use fertiliser more efficiently and are more tolerant to drought, high salts, soil compaction, heavy metals, and organic toxins.

In addition, plants are more resistant to nematodes and diseases. There have even been reports that demonstrated resistance to insects. There are 2 main groups of these important beneficial fungi. Endomycorrhizal fungi are microscopic and grow inside roots. They send out their own fungal roots and search the soil for nutrients, returning those to the plant. The plant, in turn, provides a home for the mycorrhizae, forming a symbiotic association. Endos form reproductive propagules inside roots and spores outside in the soil. manage for the mycorrhizal fungi, we could be damaging the trees.

Ectomycorrhizal relationships are common in our forests. Most trees will only form one type of mycorrhizal relationship with fungal partners. Endos are the most widespread, forming a root association with over 80% of the green plants on earth. A few plants are not endo or ecto mycorrhizal such as orchids, cabbages, beets, and mustards These fungi actually grow inside root cells and around the root and associate with most broad leaved hardwoods, fruit and nut trees together with most flowering plants, vegetables and shrubs.

Ectomycorrhizal fungi are, economically, one of the most important groups of fungi. These are the fungi that form a symbiotic relationship with a plant forming a sheath around the root tip of the plant. The fungus then forms a Hartig Net which means that there is an inward growth of hyphae which penetrates the plant root structure. There are actually seven types of mycorrhiza and 90% of plants form mycorrhiza with fungi, but ectomycorrhizal refers to this sheath forming type.

The fungus then gains carbon and other essential organic substances from the tree and in return helps the trees take up water, mineral salts and metabolites. It can also fight off parasites, predators such as nematodes and soil pathogens. Indeed, most forest trees are highly dependant on their fungal partners and in areas of poor soil, could possibly not even exist without them. Thus in forest management, if we do not manage for the mycorrhizal fungi, we could be damaging the trees. Ectomycorrhizal relationships are common in our forests. Most trees will only form one type of mycorrhizal relationship with fungal partners. Endos are the most widespread, forming a root association with over 80% of the green plants on earth.

A few plants are not endo or ecto mycorrhizal such as orchids, cabbages, beets, and mustards These fungi actually grow inside root cells and around the root and associate with most broad leaved hardwoods, fruit and nut trees together with most flowering plants, vegetables and shrubs (not laurels, rhododendrons, heather or azaleas). Vesicular-arbuscular mycorrhizas (= arbuscular mycorrhizas, VAM or AM) are associations where Zygomycete fungi in the Glomales produce arbuscules, hyphae and vesicles within roots. Spores are formed in soil or roots. These associations are defined by the presence of arbuscules.

Fungi within roots spread by linear hyphae or coiled hyphae. Vesicular-arbuscular mycorrhizas (VAM) and arbuscular mycorrhizas are mutualistic symbioses formed between the roots of most plants and fungi in the order Glomales. The Glomales are currently classified in the Zygomycetes. However, this is now being questioned. Recent information from sequences of their 18S ribosomal genes indicate that they form an ancient group branching before the Ascomycetes and the Basidiomycetes. These are the commonest types of mycorrhiza and can be found in almost all plant communities, natural and agricultural.

Vesiculararbuscular (VA) and arbuscular mycorrhizas are endomycorrhizas formed by Zygomycete like fungi and the roots of most families of Angiosperms as well as Gymnosperms, Pteridophytes and Bryophytes (liverworts). In contrast to most other types of mycorrhiza, the fungi do not substantially change the morphology of host roots. The presence of the fungi has to be determined by clearing, staining and microscopical examination of feeder roots. An experienced worker can readily recognise the coarse fungal hyphae with their distinctive angular morphology. These hyphae are restricted to the cortical region of the roots and never penetrate the stele.

All the fungal species form arbuscules—small tree like, hyphal filled, invaginations of the cortical cells—which provide intimate contact between the plasmalemmae of the two symbiotic partners and are, presumably, the point of material exchange between host and fungus. They persist in active form for a very short period, 1 to 15 days. With the exception of species of *Gigaspora* and *Scutellospora*, the fungi form vesicles within the roots. These are lipid filled, terminal swellings of hyphae with storage/perennating function. That members of two genera do not form vesicles has led to a debate as to whether the name arbuscular mycorrhizas is more appropriate than vesicular-arbuscular mycorrhizas for this relationship or whether the more appropriate of the two terms should be used depending on the fungal species present.

Suppression of Activity

Mycorrhizae function best when inorganic soluble phosphate fertiliser is used at moderate to low rates. In nature, mycorrhizae search the soil for nutrients and convert rock and organic phosphates to soluble forms and return to the plant roots. High rates of soluble phosphate fertiliser suppress the mycorrhizae. Reviewed however, mycorrhizae perform very well when used with organic fertilisers, rock phosphate, or low to moderate rates of soluble inorganic phosphate fertiliser. Likewise, certain chemical fungicides have been shown to suppress or kill mycorrhiza.

Chemicals such as benomyl, iprodione, and PCNB should not be used. In contrast, *Pythium* fungicides such as metalaxyl actually promote mycorrhizae. Insecticides and herbicides have no or little effect on mycorrhizae. Biocontrol organisms like the *Streptomyces* in Mycostop Biofungicide are encouraged by mycorrhizae. There are many scientific publications demonstrating that higher populations of beneficial organisms inhabit the root zone when mycorrhizae are present.

Mycorrhiza, Rhizobacteria and Fungi

Many soil scientists and agronomists as being the key element for cleaner, more sustainable growing systems has singled out one type of soil organism mycorrhizal fungi. "Global interest in the fungi has now reached a point whereby any discussion of agricultural biotechnology that does not include the role of endo mycorrhizal fungi is considered incomplete." Modern growing techniques using sterile growing media, sterilised or overworked chemically depleted fields, landscaping on marginal soils or sand golf greens, rely heavily of fertiliser and fungicide use to gets plants to grow. Thousands of studies and field trials show that the re introduction of mycorrhizae and symbiotic soil bacteria and fungi into these sterile conditions allow plants to flourish as nature intended, with less chemical input.

Applications of mycorrhiza

— Increased productivity
— Increased crop uniformity
— Reduced time to market
— Reduced transplant losses
— Reduced fertiliser and fungicide applications
— Increased disease resistance
— Improved crop marketability
— Accelerated growth rates
— Accelerated budding and flowering

Beneficial soil bacteria and fungi: These are fixed to an engineered mineral support, via the patented Biofixation process. These microbes drive the essential nutrient cycles of the soil, facilitate mycorrhizae, and promote the uptake of nutrients and minerals by the plant.

Membrane Effectors, Resistance Inducers, and Biostimulants: Revolutionary blends of complex plant extracts and minerals have been proven to stimulate photosynthesis and encourage the secretion of photohormones, promoting greater root, leaf, and flower growth.

Sources of Biofertilisers

Biofertilisers are organisms that enrich the nutrient quality of soil. The main sources of biofertilisers are bacteria, fungi, and cynobacteria (blue-green algae). The most striking relationship that these have with plants is symbiosis, in which the partners derive benefits from each other. Mycorrhizae are a group of fungi that include a number of types based on the different structures formed inside or outside the root.

Nitrogen is an inert gas and its uptake is possible only in fixed form, which is facilitated by the rhizobium bacteria present in the nodules of the root system. These are good biofertilisers. Endomycorrhizal fungi are microscopic and grow inside roots. Ectomycorrhizal fungi are, economically, one of the most important groups of fungi. These are the fungi that form a symbiotic relationship with a plant forming a sheath around the root tip of the plant. Vesicular-arbuscular mycorrhizasare associations where Zygomycete fungi in the Glomales produce arbuscules, hyphae and vesicles within roots.

REFERENCES

Bunders, J.; Haverkort, W.; Hiemstra, W., "Biotechnology: Building on Farmer's Knowledge," Macmillan Education, Ltd., 1996.

Diaz E (ed.)., *Microbial Biodegradation: Genomics and Molecular Biology*, 1st ed., Caister Academic Press, 2008.

Gerstein, M., "Bioinformatics Introduction," *Yale University*, Retrieved on May 8, 2007.

Martins VAP *et al.*, "Genomic Insights into Oil Biodegradation in Marine Systems", *Microbial Biodegradation: Genomics and Molecular Biology*, Caister Academic Press, 2008.

Springham, D.; Springham, G.; Moses, V.; Cape, R.E., "Biotechnology: The Science and the Business," Published 1999, Taylor & Francis. p. 1.

12

Promotion of Biofertilisers

The Government of India and the various State Governments have been promoting the nascent biofertiliser market both at the level of the user-farmer and the producer-investor through the following measures:

— farm level extension and promotion programmes,

— financial assistance to investors in setting up units,

— subsidies on sales and

— direct production in public sector and cooperative organisations and in universities and research institutions.

Over time as the industry emerges from infancy with public guidance, the following observations will be expected:

(a) increasing sales volumes and diffusion across the country,

(b) greater role of profit motivated private enterprise.

Since information on farm level usage of biofertilisers or profitability of units are not reported till date, one way to get about is by following the secondary indicators as incorporated in (a) and (b).

Biofertilisers, more commonly known as microbial inoculants, are artificially multiplied cultures of certain soil organisms that can improve soil fertility and crop productivity. Although the beneficial effects of legumes in improving soil fertility was known since ancient times and their role in biological nitrogen fixation was discovered more than a century ago, commercial exploitation of such biological processes is of recent interest and practice. The commercial history of biofertilisers began with the launch of 'Nitragin' by Nobbe and Hiltner, a laboratory culture of Rhisobia in 1895, followed by the discovery of Azotobacter and then the blue green algae and a host of other microorganisms. Azospirillum and Vesicular-Arbuscular Micorrhizae (VAM) are fairly recent discoveries.

In India the first study on legume Rhizobium symbiosis was conducted by N.V.Joshi and the first commercial production started as early as 1956. However the Ministry of Agriculture under the Ninth Plan initiated the real effort to popularise and promote the input with the setting up of the National Project on Development and Use of Biofertilisers (NPDB). Commonly explored biofertilisers in India are mentioned below along with some salient features.

Rhyzobium (RHZ): These inoculants are known for their ability to fix atmospheric nitrogen in symbiotic association with plants forming nodules in roots (stem nodules in sesabaniamrostrata). RHZ are however limited by their specificity and only certain legumes are benefited from this symbiosis.

Azotobacter (AZT): This has been found beneficial to a wide array of crops covering cereals, millets, vegetables, cotton and sugarcane. It is free living and non-symbiotic nitrogen fixing organism that also produces certain substances good for the growth of plants and antibodies that suppress many root pathogens.

Azospirillum (AZS): This is also a nitrogen-fixing micro organism beneficial for non-leguminous plants. Like AZT, the benefits transcend nitrogen enrichment through production of growth promoting substances.

Blue green Algae (BGA) and Azolla: BGA are photosynthetic nitrogen fixers and are free living. They are found in abundance in India. They too add growth-promoting substances including vitamin B12, improve the soil's aeration and water holding capacity and add to bio mass when decomposed after life cycle. Azolla is an aquatic fern found in small and shallow water bodies and in rice fields. It has symbiotic relation with BGA and can help rice or other crops through dual cropping or green manuring of soil.

Phosphate solubilising (PSB)/Mobilising biofertiliser: Phosphorus, both native in soil and applied in inorganic fertilisers becomes mostly unavailable to crops because of its low levels of mobility and solubility and its tendency to become fixed in soil. The PSB are life forms that can help in improving phosphate uptake of plants in different ways. The PSB also has the potential to make utilisation of India's abundant deposits of rock phosphates possible, much of which is not enriched.

Biofertilisers have various benefits. Besides accessing nutrients, for current intake as well as residual, different biofertilisers also provide growth-promoting factors to plants and some have been successfully facilitating composting and effective recycling of solid wastes. By controlling soil borne diseases and improving the soil health and soil properties these organisms help not only in saving, but also in effectively utilising chemical fertilisers and result in higher yield rates.

However while positive responses have been observed in a wide range of field trials, there is remarkable inconsistency in responses across crops, regions and other conditions. Even for a given crop the range of response is quite high. For example in a sample of 411 field trials carried out across districts, plant responses to inoculation with Azotobacter in irrigated wheat was observed to be significant in 342 cases and ranged from 34 to 247 Kg./Ha.

Legume inoculation by Rhizobium is the most long established practice but the responses indicated by the All India Coordinated Agronomic Research Project in the cases of mungbean, uradbean, soyabean, cowpea and groundnut all under irrigated condition were significant only in a small proportion of locations tried and failed in others. Residual effect on soil pool was not noted in most cases.

The variance of responses is similar for AZT and AZS. Dryland agriculture constitutes a very large part of agricultural area in India and also houses the majority of the poor. More than 90% of coarse cereals, 80% of groundnut and 85% of pulses come from these regions. Low productivity, unpredictable climatic swings and low dosage of chemical fertilisers also characterise agriculture in drylands. Biofertilisers, particularly Rhizobium, could be a bridge between removals and additions to soil nutrients where farmers can scarcely afford costly inputs and that too in a risky environment. But consistency in gains again eludes the trials conducted by All India Coordinated Pulse Improvement Project.

The responses usually depend on several environmental factors.

— The type of soil as measured by its water holding capacity, its levels of other nitrates, phosphate and even calcium and molybdenum (that help in protein synthesis in Rhyzobia) and the alkalinity, salinity and acidity of soil all affect the response. Higher dose of mineral nitrogen as starter suppresses nodulation, reducing response of Rhyzobium but phosphate deficiency can be an inhibitor also.

— The inadequacy of organic matter especially common in dryland agriculture is a deterrent more for the non-symbiotic strains, which essentially depend on soil organic matter for energy. Phosphobactrin response was found to be positive only in soils with high organic content and low available phosphorous.

— Soil water deficit and high temperature (hyperthermia) are prominent abiotic factors that affect nitrogen fixation in dryland agriculture.
— Native microbial population opposes the inoculants. In general predatory organisms, often already present in the soil are more adapted to the environment and out compete the inoculated population.

Apart from environmental factors, deficiencies in handling procedure are a major cause of under performance in real life application. The high sensitivity to temperature and other external conditions of these 'living' inputs, calls for enormous caution at the stage of manufacture/culture, transportation/distribution and application. This involves investment and time in research, packaging, storage and use of suitable carrier materials.

ROLE OF GOVERNMENT IN BIOFERTILISER MARKET

To attain production targets, the Government of India implemented a central sector scheme called National Project on Development and use of Biofertilisers (NPDB) during the Ninth Plan for the production, distribution and promotion of biofertilisers. A National Biofertiliser Development Centre was established at Ghaziabad as a subordinate office of the Department of Agriculture and Cooperation with six regional centres. The purpose of the scheme covered organisation of training courses for extension workers and field demonstrations and providing quality control services. Production and distribution of different biofertilisers were also undertaken but subsequently discontinued as the centres redefined their role towards R&D and HRD related activities. Capacity creation and production was however encouraged through one time grant for new units.

The financial assistance, given as grant-in-aid to the tune of Rs 13 lakh and now increased to Rs 20 lakh per unit and thrown open for all, was routed through the State governments but owing to delays in release of grants the onus is transferred to NABARD/NCDC. The public sector organisations form a bulk of the units in the industry, while similar units in the private sector are also coming forward. Different State governments also provide subsidies sometimes up to 50% of the sales realisation but the manner of subsidisation is rather unsystematic. In many cases the discrimination and manipulation in subsidising lead to a lot of intra industry variation in prices.

The government also plays a dominant part in marketing biofertilisers in three possible channels:

— State government via District level Officers and Village level workers to farmers,

— State Marketing federation via cooperative bodies to farmers and

— State Agro-industries Corporations via Agro service Center to farmers.

The producers are however free to sell through their own sales network or through market.

The Fertiliser Association of India (FAI) periodically presents information compiled on capacity and distribution of biofertilisers by various units. In the absence of reported information on farm level use of the inputs, this can help in understanding the progress of the technology and its adoption in India. The period covered by the data is 1992-93 to 1998-99.

The FAI report gives the distributions of different strains for recent years by states that can proxy for usage by farmers. For a better understanding of the demand for use, firm level information on capacity, distribution and prices would be more useful. However the FAI could not report for all existing producing units due to their non-responses and this irregularity is more for distribution and prices. So the inferences drawn in the present study are only based on the samples that report the required information. The FAI reports give information of annual distribution levels of various inoculants and their sale prices for consecutive years by firms.

BIOFERTILISER TECHNOLOGY

Government of India and the different State Governments have been promoting use of biofertilisers through grants, extension and subsidies on sales with varying degrees of emphasis. With time farmers too learn about the technology forming their perception on the basis of agronomic realities of their regions, the knowledge gained from experiences of farmers around them and including themselves and the information provided by different disseminating agents and form their own decisions of adoption. Above all the enterprise of the firms working through their marketing, research and development efforts would lead to the widespread use of the inputs once the prospect of profit is sensed. Commercial appeal with the passage of time and government support.

Progress of the Industry

Based on the data for 1995, 1997 and 1999 it appears that the industry witnessed a steady increase in the number of units producing the input. Over the period of four years the number of units went up by 53% from 62 to 95 and further to 122 in 2002 (Ministry of Agriculture, GOI). The total capacity expanded by 102% going by the information on units reporting their capacities. New private units joined the industry improving their numeric share while the public sector, after the initial burst slowed its pace. However, a deeper look would be more illuminative.

The total distribution as reported by the units on an annual basis increased over time at an impressive rate of over 50%. However it is clear that the bulk of the growth took place by 1992-95 of the sample period and stagnated thereafter. There are also changes in shares by types with moderate success in AZT and by far the best performance by PSB. The decline in RHZ indicates success in groundnut and pulses was below expectation.

A measure of capacity utilised is obtained relating actual distribution (as opposed to production) to capacity. The industry has been going through an adjustment of size as average capacity of a unit came down from 261.8 tonnes to 205.6. The capacity addition in the industry was less relative the addition of new units due to entry of lower sized new units. The average distribution also declined in the first two years possibly signalling the need for a down size and picked up subsequently. The average capacity utilisation has been poor but the down sizing may have arrested the declining trend.

The chemical based fertiliser technology incorporated in the green revolution was successful by its rapid adoption rate but the unbalanced spread across the country, especially in the eastern region marks a crucial failure. The central government's role in the new biofertiliser technology would be justified by greater spatial dimension of the success.

Table 1 : Average Capacity, Distribution and Capacity utilisation of Units

Year	*Capacity (Tonnes)*	*Distribution (Tonnes)*	*Capacity Utilisation*
1994-95	261.8	111.3	0.43
1996-97	225.8	87.91	0.39
1998-99	205.6	94.37	0.46

Since biofertilisers are perishable and sensitive to quality of handling, the distribution of plants would to some extent reflect the regional distribution pattern. However this is only partially valid as units with large distribution networks do distribute over larger areas. Located in Phulpur in Uttar Pradesh, IFFCO's MLN Farmers' Training

Institute produces all strains of biofertilisers and have distributed in states other than the home state. Eastern states like Bihar and Orissa are also served though the share has gone down notably for West Bengal where distributions came down to nil in 2000-02.

Table 2 shows how the industry started with a fairly even dispersion of units gave way to a concentrated locational pattern. While both east and north lost their shares, there was continuous and significant expansion in the western region. The share of south remained stable. Some of the states saw closure of units and in fact over the period the number of units in north actually came down.

Table 2: Regional distribution of Bio-fertiliser Plants (%)

	1995	*1997*	*1999*
Region			
East	17.74	13.89	12.63
North	29.03	27.78	17.89
West	22.58	26.39	38.95
South	30.65	31.94	30.53

The distribution does not follow that of chemical fertilisers they supplement, where north is the largest claimant. However the eastern region comes last in share in both chemical and biofertilisers though the share in the latter case is even more diminutive. The comparison however is not complete without bringing in the cropped area in the regions.

Influence of Private Enterprise

In the nineties India stepped a market oriented regime and the desired role of the government became one of facilitator. In the past public sector units including state owned cooperative organisations played a crucial role in implementing government's social objectives on a continuing basis, compromising commercial interests in the process. In a more competitive environment failure to attain commercial viability does not augur well for the sustenance of the organisations as also for the public budget.

In the case of biofertilisers too the initiative taken by the public sector along with numerous universities and research units that are also state funded must with time lead to commercial success once the technology is transmitted to the field and this in turn is expected to draw private enterprise since the market is open for entry. The government is also encouraging private entry by extending financial grant to all and sundry. Table 3 suggested that slow down in private commercial sector distributions is more pronounced than total.

Table 3 : Use intensity of Biofertilisers (BF) and Chemical fertilisers in Indian Agriculture

Region Biofertilisers		Chemical (NPK)	Cropped area	Chemical (NP)
	(Kg/hect.)	(%)	(Kg/hect.)	(Kg/hect.)
South	125.21	18.46	107.06	0.05
North	130.43	22.32	127.65	0.01
West	60.82	40.91	57.07	0.06
East	70.63	18.32	61.73	0.01
Total	90.04	100.00	82.90	0.04

Although many of the new entries between 1995 and 1999 are private commercial enterprises, their share in total capacity came down steadily as revealed by the reports (Table 4). This is because of scaling down of size from a relatively high level of 504 tonnes to 300 tonnes as already noted. The coexistence of smaller new units with the larger ones of higher vintage has increased the variety in industry as measured by the coefficient of variation. The share in distribution however has been relatively more stable despite showing a slight declining trend in the last five years reported. Over the years the industry as a whole distributed about 35.6 thousand tones of which a little less than half is accounted by the commercial private firms. The latter also charged higher prices on the average and there is slight rise in prices between 1995 and 1998.

Table 4: Capacity of Privately owned Units

	1995	*1997*	*1999*
Average Capacity (tonnes)	504	437	308
Coefficient variation (%)	120	143	162
Share in total capacity (%)	74	69	67

Structure of Industry

Based on the 50 units that reported capacity, distribution and prices in 1998-99. An econometric analysis will attempt to use this information to explain the distribution performance by the structural components of the industry. The capacity of a unit is defined by the various facilities of production including equipment for various operations, infrastructure and space. Due to the complex nature of the process involving laboratory culture of life forms, that requires definite combinations of space, equipment and time, there is little substitutability among the inputs.

Numerous studies on technological evolution emphasized the developmental role of a firm and the strength of its sales network, creating market and drawing market feedback, for its success. In general, firms with larger production facilities are expected to invest more on networks to understand and access the market but it is not uncommon for firms with larger distribution networks to act as marketing agents for smaller units who are lacking and in few rare cases like that of NAFED the distribution even exceeds capacity. The sales networking would be stronger also for concerns that are in some way already in the business of selling agricultural inputs. Since the exact scope and nature of the units or possibly their parent companies is not clear from the data, past experience in selling biofertilisers may be considered as an indicator of their marketing capabilities.

Judging by the sample information, units that record a cumulative distribution of more than 100 tonnes over the previous 3 years are deemed to be enjoying greater selling experience and broad based net-workings. While the cumulative distribution performance takes care of 'le arning by doing' opportunity of the firm a possible addit-ional characteristic could be the age or vintage of the unit, which also allows for 'learning by looking' opportunity. Since data on vintage is not readily available, those that existed and reported for March 1995 are classified as older units (VINT). Units that produce both nitrogen fixers and phosphate solubilisers are categorised as joint producers (JOINT).

Production of biofertiliser started in India with significant government involvement with active participation of the public sector that is directed more by public policy and social objectives than profit. The extent of commercial success would be indicated by the participation of private commercial units so long as market is free for entry. The private firms that reported can be categorised as private (PVT) and others that include universities, research institutes, cooperative, agro-marketing and other public sector organisations. Finally, the regional distribution of the firms is of interest in view of the tendency of concentration observed.

Going by a size categorisation of units based on sample average capacity of the units as the cut off size, less than 30% of units are classed as large, with majority being relatively small units. The share of private sector is much larger (64%) for the smaller units than the larger ones (36%). More than 70% of the large units are of longer vintage reiterating that new entry has been mostly in small units. About 70% of the small units came into being after March 1995. The small units show some tendency to specialise in either nitrogen fixers or phosphate solubilisers while all the large units produce both kinds. Both size categories show regional bias for west, followed by south, though smaller units have relatively greater presence in south as also in east where there is no presence of large units. The average capacity differs widely between the two categories, as does employment generation but large units record higher capacity utilisation. Average price charged is marginally in large units.

The following equation is estimated to explain distribution:

$$Dj = a_0 + a_1\, P_j + a_2\, CAPACITY_j + Oa_{3k}\, a_k$$

Where D is distribution in tonnes in the completed year, P is sale price in Rupees per kilogram, CAPA is capacity in tonnes at the year end and a are k dummy variables corresponding to unit Price, calculated as weighted average price of different biofertilisers (PBIO) with distribution shares as weights, varies moderately among the units. As a deflator the price (PCHEM) of chemical fertilisers, N and P, are taken for the year and weighted by consumption shares in the state concerned. The deflated price variable P-DEF is expected to show a negative effect so long as there is some competition between chemical and biofertilisers.

A specification including both prices separately is also estimated. CAPACITY of the unit accounts for scale of the unit and the dummy variables place the units in different categories that could have a bearing on distribution. Variable DUM-EXP is used to capture units with selling network as indicated by past experience as discussed. Another variable for experience is VINT. Dummies WEST, EAST and SOUTH delineate the units by three major regions to see if any one region makes a difference over all others including north. Variable PVT is used to examine the performance of units in private commercial sector. Similarly, variable JOINT helps find out if varied production pattern is an added advantage.

The equations coming with reasonably good fit show expected signs of price and scale variables though there is no significant effect of price. The scale variable CAPACITY has a positive and significant coefficient. Firms with greater selling experience (DUM-EXP) also distribute significantly higher amounts given other conditions unchanged. Variable PVT comes with a negative coefficient suggesting that private commercial units distribute less than others after controlling for price capacity and experience.

The region dummies too peculiarly indicate the near absence of regional dimension in the distribution after controlling for relevant factors. Even units located in the western region, which claimed the maximum expansion of the industry, shows no remarkable advantage over others and the variable comes with an insignificant and even a negative coefficient. Joint production of different biofertilisers gives no added advantage over specialised units and older units shown by variable VINT do distribute more, possibly gaining from being in the field for long.

The analysis conducted with the limited data available throw some light on the progress of the technology through the indirect indicators from industry. The empirical findings can be summarised as follows:

- First, the distribution of biofertilisers, proxying for its adoption rate has not consistently grown over time and has slowed down in the late nineties. Starting from a small base one would have expected a faster and possibly accelerating growth performance as the input finds greater acceptance.
- Second, although there have been more and more new entries in the market, the average capacity came down characterising the industry by a large number of small

units. While size adjustment in infant industry is normal, it must be borne in mind that distribution of an agro input also calls for substantial sales networking and a deep understanding of the field reality in agriculture.

— Third, there has practically been no diffusion of the technology despite the central government's interventions and the distribution among units has tended towards greater concentration especially in Maharashtra and other states of the west and south. As with chemicals, the impact on the east has been poor. Possibly the interventions and the policy of the State governments proved more decisive and in some cases, the recommendations of the agricultural universities or lack of them (as in Punjab) had a role in the farmers' lack of interest in the technology.

— Fourth, despite entry of private players, the share of the private commercial sector in distribution remains below 50% and over time the private firms have neith improved their share in capacity or distribution n their growth rate of distribution. This casts doubt o whether the business is as of now viewed as commercially gainful prospect.

— Fifth the regression analysis further brings out some aspects of concern. There seems to be little intrinsic justification for the regional preferences shown by units since the distribution performance does not appear better in any region, even the west, than others after controlling for unit size and price.

Government Intervention

The result of the analysis is not positive with respect to government intervention in the market is concerned. Till date the central government has spent several crores of rupees as grants to invite investment and state governments have also spent large sums for subsidising and promoting use. The questions that now arise are (a) Does the government need to intervene and support the market to build up? (b) If so what should be nature of intervention? Despite the lack of expected response there still is a strong case for government to intervene and possibly subsidise in the market. The reasons can be broadly classified in two groups:

If biofertilisers impart certain social and long-term gains for which private individuals may not be willing to pay at least until the gains become 'visible', there is a rationale for spreading the cost over a larger group of beneficiaries or the society at large. The government can act on behalf of the society through appropriate policy even if the market is otherwise competitive.

1. Biofertilisers have important environmental and long-term implications, negating the adverse effects of chemicals. At the farm level, the gains from increased use of the technology can spill over to other farms and sectors through lesser water pollution than chemical fertilisers and even to an extent organic manures can create.

   The gains from the new technology coming through the arrest of soil damage may not be perceived over a short span of time unlike for chemical fertilisers, which yield quick returns. At the same time the farmer has to incur considerable initial cost in terms of skill acquisition, trial and failure and risk. In agrarian situations where agents often operate with bounded rationality, adoption may be slow and influenced greatly by neighbours' experiences over time. Empirical evidences in agriculture around the world show that adoption of new practices take time to pick up and the earliest adapters are often conservative about the percent of acreage devoted to the new technology. The state has a role to play in inducing farmers to adopt improved practices.

3. The producer firms have serious uncertainty about the demand or saleability of the product, which deters investment, particularly if it is irreversible. The success or failures of early entrants who take the initiative or those who indulge in research for an improved product convey important information to others and thereby to society. The market however does not always reward the initiative. The capital market is also not always ready to provide the risk capital at reasonable rates.

In many cases the market itself is not competitive and although the first best solution would be to promote competition, there are other ways of dealing with the situation.

1. Although biofertilisers have been promoted as supplement or complement of chemical fertilisers, in reality they are two alternative means of accessing plant nutrients. The strength of complementarity as against substitution between the two inputs is open to empirical verification, but there is no denying that farmers and producers do perceive the substitutability relation to an extent. The pricing of chemical fertilisers is far from marginal cost based. In particular urea is under administrative pricing and there are serious economic and political compulsions to continue the protection although the movement in spirit has been towards openness. In such circumstance the price of biofertilisers along with the risk and responses will be weighed with those of chemical fertilisers, and promotion of the technology for environmental reasons would call for some degree of protection to minimise the inter-fertiliser price distortion.
2. The external or environmental cost of using chemical fertilisers, though not measurable may also be taken into account when comparing with biofertilisers if the latter is to be promoted.

The next question that arises is what should be the nature of intervention. With the results at hand, the central government needs to improve its strategy towards a more broad-based and integrative role in order to provide the correct environment for the adoption of the technology.

One of the main barriers faced by the producers and investors is inadequate demand and the inconsistent and seasonal nature of the existing demand. It may be recalled that the technology is as yet nascent and evolving. The rice dominated eastern region remains a nonstarter and the wheat-rice growing north has not shown much interest either. Research on developing efficient, temperature tolerant and hardy strains is a vital step to the actual success of the technology. In particular, there can be some focus on the potentials of the technology in rice and cereals in general although its significance for crop diversification is of equal concern.

The government has a dominant role to play in encouraging and funding research either by universities or private firms to make up for shortfall in private commercial research initiative. As is well recognised now, in a market-oriented system, information is often scarce and costly and hinders many business activities. For biofertilisers, the government has a dominant role of disseminating the technology through demonstration, training and other activities. A proposal to make it mandatory for biofertiliser units to contribute in the process through their own demonstration is a welcome step. Research and extension/ promotion both must be to the extent possible specific to local conditions and constraints.

It is a good practice to promote biofertilisers as an input conjunctive to other forms of fertilisers, but keeping in view the protection given to chemicals, there is some ground for subsidising the former to encourage the ir use. However, there is a need to work out a systematic and uniform way to give out subsidies so that they do not distort inter unit prices and help some units at the cost of others. The States should be strongly guided on this norm. The main purpose of the subsidies would however be to induce farmers to try out the input at affordable and acceptable prices rather than to support certain producers directly. As far as producers are concerned, a healthy competition would only help develop a market in the long run and benefit the farmers.

Open sales in the market may be encouraged to improve the capability of the units in marketing products. Any attempt to fix a minimum price for the sake of quality or other consideration will only go against the spirit of competition and harm the interests of the industry and the farmers. The present study finds some but not conclusive role of the price factor but scientific studies may be conducted on the farmers' 'willingness to pay' for the new input to determine the need and extent of subsidies.

The subsidy, if it is considered desirable, could be routed through the Ministry of Environment since the main aim of the policy is to encourage environmentally improved practices. Quality is an essential element in this market and the entry of new units also heightens the threat of poor strains appearing in the market and ruining the farmers' confidence in the technology and their incentive to adopt the same. There is a strong case of quality control and legal safety nets, which could possibly draw the involvement of local bodies backed by the expertise of specialised agencies.

Since the microorganisms are extremely perishable and sensitive to local conditions and handling the farmer always encounters some risk in opting for the input. To the extent the farmers' risk perception is responsible for inadequate adoption there is a case for spreading the risk over a larger society. An insurance or a buy back scheme could work with the cost shared by the government and the producers but this would call for some monitoring of farmers' behaviour. Such a scheme would also reduce the need for subsidies.

The grant in aid given for new units and possibly in future for upgrading older units is not really a desirable method. To the extent that inadequate demand is the main hindrance to producers, the financial aid would not help at all. In fact the method embodies an incentive for *moral hazard* where the firm receives the grant but does make enough 'effort' to sell. The transfer of the responsibility to NABARD would only impose undue burden on the bank in monitoring sales attempts.

The regression analysis says that even with the same capacity build up (with the help of grant), a private commercial unit sells less than others. In actuality, business organisations have their own dynamism and sales effort would definitely follow from commercial prospects, which can come from greater acceptability on the users' front. Concessional finance could be arranged for investment to encourage investment as well as efficiency in follow up business.

The scaling down of average size of unit must be associated with a review of sales and development capability of units. It is noted that smaller units do not necessarily enjoy higher capacity utilisation in terms of distribution. The broader networks, deeper understanding of market and greater sales experience of older and larger units, specifically the cooperative sector units involved in marketing other agro inputs may be commercially exploited for the marketing of products produced by smaller and less experienced units and the development of a synergic system in this regard may be useful.

REFERENCES

Ange, A. L., *Integrated plant nutrition system* on farm and soil fertility management: The *major concepts*, Rome: FAO, 1992.

lston. J. M., P. G. Pardey, and V. H. Smith, "Financing agricultural R&D in rich countries: What's happening and why", *Australian Journal of Agricultural and Resource* 42(1), 1998.

Bibliography

American Society of Agronomy, *Decision reached on sustainable ag.*, Agronomy News, January, pg 15, Madison, Wisconsin, 1989.

Bartholomew, G.A., *The roles of physiology and behavior in the maintenance of homeostasis in the desert environment*, Homeostasis and Feedback Mechanisms, Symposia of the Society for Experimental Biology 18:7-29, 1964.

Belimov, A. A., Kojemiakov, A. P. & Chuvarliyeva, C. V., "Interaction between barley and mixed cultures of nitrogen fixing and phosphate-solubilising bacteria," *Plant and Soil*, 173, 29-37, 1995a.

Bockman, O. C., O. Kaarstad, O. H. Lie, and I. Richards, *Agriculture and fertilizers: Fertilizers in perspective*, Oslo: Norsk Hydro, 1990.

Conway, G. R., and J. N. Pretty, *Unwelcome harvest: Agriculture and pollution*, London: Earthscan Publications Ltd., 1991.

Coster, R., "Alleviating fertilizer supply constraints in West Africa", In *Alleviating soil fertility constraints to increased crop production in West Africa*, ed. A. Uzo Mokwunye, Dordrecht: Kluwer Academic Publishers, 1991.

Cox, P.G., N.D. MacLeod, and A.D. Shulman, *Putting sustainability into practice in agricultural research for development*, Submitted to the United Kingdom Systems Society 5th International Conference on Systems for Sustainability: People, Organizations and Environments, 7-11 July, De Montfort University and The Open University, Milton Keynes, 1997.

Donovan, G., and F. Casey, *Soil fertility management in Sub-Saharan Africa*, World Bank Technical Paper No. 408. Washington, DC: World Bank,. 1998.

Dregne, H. E., "Erosion and soil productivity in Africa", *Journal of Soil and Water Conservation* 45:431-436, 1990.

Freney, J. R., "Efficient use of fertilizer nitrogen by crops", In *Appropriate use of fertilizers in Asia and the Pacific*, ed. S. Ahmed. Taipei: Food and Fertilizer Technology Center, 1996.

Gill, G. J., *Major natural resource management concerns in South Asia*, Food, Agriculture, and the Environment Discussion Paper 8. Washington, DC: IFPRI, 1995.

Ito, O and Watanabe, I., "Availability to rice plants of nitrogen fixed by *Azolla*," *Soil Sci. Plant Nutr.*, 31 (1) 91-104, 1985.

Janes, R., "Growth and survival of *Azolla filiculoides* in Britain I. vegetative reproduction," *New Phytol.* 138, 367-375, 1998a.

Khaosaad, T., Garcia-Garrido, J. M., Steinkellner, S. & Vierheilig, H., "Take-all disease is systemically reduced in roots of mycorrhizal barley plants," *Soil Biology and Biochemistry*, 39, 727-734, 2007.

Kondo, M, Kobayashi, M and Takahashi, E., "Effect of *Azolla* and its utilisation in rice culture in Niger," *Plant and Soil* 120, 165-170.

Ram, H. and Prasad, J., "Comparative study of *Azolla* and N, P, K fertilisers on wheat crop," *Agric. Res. Rural Dev.* 5, 43-45.

Rengel, Z. G. D. Batten, and D. E. Crowley., "Agronomic approaches for improving the micronutrient density in edible portions of field crops," *Field crops research* 60: 27-40, 1999.

Shenker, M., I. Oliver, M. Helmann, and Y. Hadar., "Utilization by tomatoes of iron mediated by a siderophore produced by Rhizopus arrhizus," *J. Plant Nutr.* 15, 2173-2182, 1992.

Singh, S. & Kapoor, K. K., "Inoculation with phosphate-solubilising microorganisms and a vesicular-arbuscular mycorrhizal fungus improves dry matter yield and nutrient uptake by wheat grown in sandy soil," *Biology and Fertility of Soils*, 28, 139-144, 1999.